Pocahontas

Ætatis suæ 21. Æ.
1610.

Pocahontas

Medicine Woman, Spy,
Entrepreneur, Diplomat

PAULA GUNN ALLEN

HarperSanFrancisco
A Division of HarperCollins*Publishers*

POCAHONTAS: *Medicine Woman, Spy, Entrepreneur, Diplomat.* Copyright © 2003 by Paula Gunn Allen. All rights reserved. Printed in the United States of America. No part of this book may be used or reproduced in any manner whatsoever without written permission except in the case of brief quotations embodied in critical articles and reviews. For information address HarperCollins Publishers, Inc., 10 East 53rd Street, New York, NY 10022.

HarperCollins books may be purchased for educational, business, or sales promotional use. For information please write: Special Markets Department, HarperCollins Publishers, Inc., 10 East 53rd Street, New York, NY 10022.

HarperCollins Web site: http://www.harpercollins.com
HarperCollins®, ▇®, and HarperSanFrancisco™ are
trademarks of HarperCollins Publishers, Inc.

FIRST HARPERCOLLINS PAPERBACK EDITION
PUBLISHED IN 2004
Designed by Joseph Rutt

Library of Congress Cataloging-in-Publication Data
Allen, Paula Gunn.
Pocahontas : medicine woman, spy, entrepreneur, diplomat /
Paula Gunn Allen. — 1st ed.
p. cm.
Includes bibliographical references and index.
ISBN 0–06–073060–9 (pbk.)
1. Pocahontas, d. 1617. 2. Powhatan women—Biography. 3. Powhatan women—
History—Sources. 4. Smith, John, 1580–1631. 5. Rolfe, John, 1585–1622.
6. Jamestown (Va.)—History. I. Title.
E99.P85P57145 2003
975.5'01'092—dc21
[B]   2003056583

04 05 06 07 08 RRD(H) 10 9 8 7 6 5 4 3 2 1

Dedication

To my great-grandmother Meta Atseye Gunn, graduate of Carlisle Indian School, who, they said, walked in balance between two worlds, American and Laguna Pueblo.

To all the mixed-culture descendants of Grandmother Pocahontas. She was the first boarding-school Indian, and the first to walk two paths in a balanced manner. She was the mother of Thomas, of Capahowasc, of Virginia, and of the American *powa,* "dream." The *powa* she carried has not died: it has flourished. It is woven of two visions of two hemispheres; it is the tradition of *powatan,* we who dream with one another. May the multiheritage *powa* of the world make all people whole.

Da-wa'e.

Everyone must die. It is enough that the child lives.
—Lady Rebecca Rolfe, "Pocahontas"

Contents

Acknowledgments

I would like to thank the many peerless people who offered help, support, and inspiration. Thanks go to Cara Marianna, who read and sorted out the most confusing chapter of the manuscript and listened to me, offered advice, gave guidance, and provided great commentary; to Charlotte Gullick, who painstakingly read the entire manuscript, for her encouragement, careful commentary, and frequent acts of kindness; to Mary Churchill, student, friend and colleague, for her invaluable research and advice; to Hunter, for her fine gift of the King James dime and the simplified map of the *tsenacommacah;* to Melinda Faye, for her flattering photo of me; to Rosie Bock, for reading and providing thoughtful commentary; to Renee Crowley, for reading and running off reams of hard copy; and to my editor at Harper San Francisco, Eric Brandt, for persistence, courage, and a clear hand and eye.

Particular thanks go to my children: Lauralee Brown Hannes, for words of wisdom, truth, and balance when most needed; Sulieman Russell, for inspiration, stimulating conversation, helping me over writer's block(s), and research; and Gene Brown, for continuing inspiration throughout this long process. His death acted as a spur, his life as a guide. Others who helped and blessed the work along the way with their conversation and knowledge include Sandi Pineault and Melissa Tantaquidgeon (Mohegan); Jack Forbes (Delaware); James Axtell, Charles de Lint, F. David Peat, Betty Donahue Chocktaw, and Michelene Pesantubbee (Chocktaw); and Patricia Clark Smith (Micmac). Special thanks to the Gaea Foundation of Washington, D.C., who hosted me to two months in Provincetown, where I could soak up Algonquin whispers. Thanks also go to the Wampanoag women and men at the Plimouth Plantation Wampanoag exhibit, for their warmth, guidance, conversation, and hard work to re-create life in a sixteenth-century Wampanoag village; to the women at the visitors' center at Gravesend, England, for help and a pleasant chat; and to Suzanne Byerly, for her classes-cum-community where a writer can be, write, listen, and receive that particular kind of food we so need.

I would especially like to acknowledge the manito and Lady Rebecca, "Pocahontas," for their continuing help, direct and startling messages, and shared courage.

Chronology

Tsenacommacah Chronology

Matoaka/Rebecca (Pocahontas was her childhood nickname) was probably Mattaponi/Pamunkey, both tribes that were part of the Powhatan Alliance (known to its members as the *tsenacommacah*). Dates as Westerners count them—or estimations where indicated and possible—are approximate at best until 1608; earlier ones are added to this chronology as a guide rather than as fact. Historians agree that Matoaka (Pocahontas) was born sometime around 1597 and died in 1617. She was thought to be around twenty-one years old when she died. The four hundredth anniversary of her fateful meeting with the English arrives in January 2009. The formal events surrounding the landing at Jamestown will begin in 2007.

MATRIX

ca. 1500	Prophecy foretells end of the *tsenacommacah*
ca. 1570–71	Wahunsenacawh taken hostage by Spaniards
est. 1589	Wahunsenacawh returns to the *tsenacommacah*
by 1595 or 1596	Wahunsenacawh *mamanantowick* of Tsenacommacah
	Matchacómoco (Grand Council) learns prophecy of Spirit Woman to be born

CHILDHOOD

ca. 1596	Matoaka born; signs say she is the woman foretold
ca. 1596	Matoaka gets her child nickname, Pocahontas
est. 1603	Pocahontas given her Dream-Vision
est. 1603	Pocahontas taken to Werowocomoco
1607	Virginia Company comes to the *tsenacommacah*
1608	Pocahontas becomes spiritual guide to Nantaquod (John Smith)
1608	Pocahontas visits James Fort
1608	Pocahontas fetes Nantaquod (Smith) and his men
1609	Pocahontas conveys message to Nantaquod (Smith)

WOMANHOOD

1610	Pocahontas marries Kuocum
1611	Pocahontas gives birth
1612	Pocahontas goes to James Fort
1613	Pocahontas gets remade; becomes Lady Rebecca
1614	Lady Rebecca/Pocahontas and John Rolfe marry; settle at Varina
1614	The Peace of Pocahontas begins
1614–16	Lady Rebecca/Pocahontas instructs Rolfe in the ways of *manito apook*
1615	Lady Rebecca and John have first child, name him Thomas

1616	Lady Rebecca and John, along with her retinue, go to England
1616–17	Lady Rebecca introduced into high society in England
1617	Lady Rebecca dies and is buried at Gravesend, England

AFTERMATH

1618	Wahunsenacawh dies
ca. 1619	Ceremony of World Renewal complete
	(Tobacco becomes mainstay of Virginia Colony economy)
1622	The Peace of Pocahontas ends

English Chronology

There are dates in the English time line that have direct bearing on the story of Pocahontas's life. Major events are listed here. The Roanoke Colony was established in 1585 and ended sometime before 1587. The English came to the *tsenacommacah* in 1608, English time. I have listed in order the events that had direct bearing on her life and her destiny as I have given them in this work.

1558–1603	Elizabeth I, ruler of England, Ireland, and Wales
1584	Virginia Colony letter of patent granted to Walter Ralegh
1567	James VII crowned king of Scotland and France

1603	James I crowned king of Great Britain
1589	James VII and Anne of Denmark wed
1585–87	Roanoke Colony (Virginia)
1606	Virginia Company receives new crown patent from James I
1607	English begin Jamestown Colony settlement
1608	Smith captured by Chickahominy
1609	John Rolfe arrives at James Fort
1622	John Rolfe dies
1619	Plantation system based on tobacco production firmly established
1625	James I dies

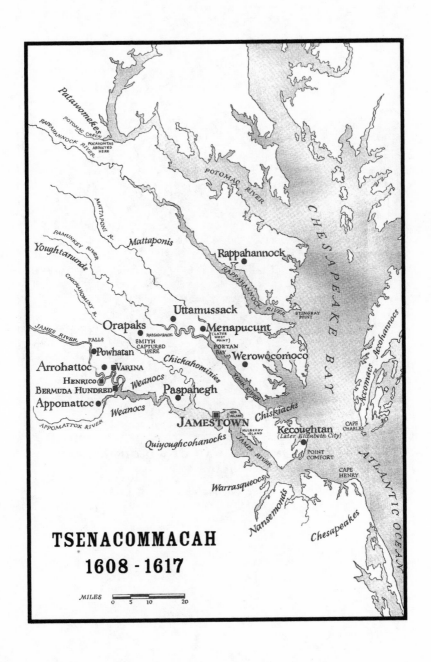

Patawomekes

POTOMAC CREEK

RAPPAHANNOCK RIVER

POCAHONTAS ABDUCTED HERE

POTOMAC RIVER

MATTAPONI R.

PAMUNKEY RIVER

Youghtanunds

CHICKAHOMINY R.

Mattaponis

Rappahannock

RAPPAHANNOCK RIVER

STINGRAY POINT

Uttamussack

JAMES RIVER

FALLS

Orapaks

RASSAWACK

SMITH CAPTURED HERE

Menapucunt
(LATER WEST POINT)

POETAN BAY

Powhatan

Chickahominies

Werowocomoco

YORK RIVER

Arrohattoc

Varina

HENRICO

Weanocs

BERMUDA HUNDRED

Appomattoc

Weanocs

Paspahegh

APPOMATTOX RIVER

JAMESTOWN

HOG ISLAND

Chiskiacks

MULBERRY ISLAND

Kecoughtan
(Later Elizabeth City)

CAPE CHARLES

Quiyoughcohanocks

JAMES RIVER

POINT COMFORT

CAPE HENRY

Warrasqueocs

Nansemonds

Chesapeakes

CHESAPEAKE BAY

Accomacs

Accohannocs

ATLANTIC OCEAN

TSENACOMMACAH
1608 - 1617

MILES   0   5   10   20

JOHN SMITH'S MAP OF VIRGINIA
(1612)
(courtesy of the Virginia Historical Society)

Oo-maa'o / Introduction

We need to remember that because of both the historical
documents and the fluid nature of a life so integrally
connected to the *manito aki,* we will never entirely know
Pocahontas.

—Charlotte Gullick

Who has said we come to earth to live
We only come to dance
We only come to dream

—Nahuatl

P ocahontas: *Medicine Woman, Spy, Entrepreneur, Diplomat* is an
American Indian story, as Pocahontas was an American Indian
woman. It must be told in a way that is faithful to the traditional way
of telling our life stories. Pocahontas, like her people, can never be
known in terms of "facts," bereft of the spiritual tradition that defined
her and her people, or understood outside the spirit-centered world
they inhabited. A biography of Pocahontas must tell her life in terms
of the myths, the spirits, the supernaturals, and the worldview that informed her actions and character. The modern way to write a biography involves, among other things, a process that singles out an
individual, cuts her out of the total biota, or life system, within which
she lives and from which she derives her identity, and gives her value
and prestige *above* the rest. But in Native traditional life stories, the
subject of the biography—or, often, autobiography—is situated
within the entire life system: that community of living things, geog-
raphy, climate, spirit people, and supernaturals. In this mode, the sacred narratives, ceremonial occasions, daily concepts and assumptions,
and interactions with the other world are included as a matter of
course. For it is only within the narrative tradition of a community
that an individual member of that community can be recognized.
What is emphasized is the interactive nature of that person's life, as a
given. Thus personhood is not seen as a product of economic forces,
politics, divine providence, or any forces that operate entirely outside
that person. This means that traditional biographies make privileging
the subject as a victim or oppressor, sinner or saint, impossible. A life
comprises all the currents that flow through it and that it flows
through. Nor is it ever finished. A person, like life itself, has neither
beginning nor end. It just is, here and there, self and other, present
and absent. In that way, Pocahontas's life continues on, as the books,
poetry, songs, dramas, and films centering on her prove.

It has also long been the modern way to frame the female as an
adjunct to the male. In this mode, a man is a person. A woman is attached to a male, thus enjoying a sort of vicarious personhood. One
locates a woman's identity by showing her relationship to the real
human being—husband, father, brother, or king—and thereby judges

her significance. This convention, like the separation of the person from the matrix of her life, is not a convention of the Oral Tradition in Indian country. It may seem to be, because biographies, like other narratives, are a mode of Western discourse, and are almost always written by Westerners. On a few occasions they are written by Native historians, biographers, or journalists, but since academic publishing and advanced-degree granting depend on the writer's mastery of Western modes of discourse, the point is moot as yet. However, biographies known as "as told to" biographies, in which the subject tells of her or his life and this telling is recorded by someone who then transcribes it and readies it for publication, serve as examples of how to write a biography of a Native person and avoid the distortions that occur when such a life is transferred into a Western narrative context.

Pocahontas: Medicine Woman, Spy, Entrepreneur, Diplomat is not exactly an exception to either narrative order. While I am constrained and guided by Western conventions for biography, I have made every attempt to "Nativize" this work by structuring it to conform as much as possible to the narrative conventions of American Indian Oral Traditions. I have endeavored, as much as possible, to present the life of Matoaka, who is more familiarly known by her nickname, Pocahontas, in the Algonquin Indian context in which it was lived. I have incorporated several conventional rhetorical devices into this retelling of the life of Pocahontas. One is the use of reiteration, usually referred to as *repetition*. While not a device favored in modern nonfiction, it is a ubiquitous element in narratives, chants, and songs from the Indian literary tradition, whatever tribe or Native Nation. I have changed each iteration slightly, adding more information, or shifting the point of view somewhat, in an endeavor to suggest the Native convention, to help readers with more new information than is usual in a biography, and to act as a textual reminder that we're not in Kansas here, but in the *tsenacommacah,* the communal lands and being, of the Powhatan people. As I have used a number of Algonquin concepts and words, I have included a glossary; I have also always italicized such words unless they are proper nouns—the only exception to this being *manito,* as it is sufficiently part of the American lexicon to count as an American word. I have also refrained from capitalizing Algonquin words unless they are proper names or appear at the beginning of a sentence or in a title—a practice that is common among Native writers and scholars.

Because Pocahontas was not of our period or of our modern American culture, I have chosen to base my interpretations of her actions on a Powhatan worldview. Thus, I am basing my narrative on several assumptions: that manito—a complicated word that relates to paranormal, supernatural, and transcendent conditions of consciousness, existence, and event—is reality. I am also assuming the reality of the *manito aki,* the spirit world or realm where supernaturals live and where the laws of physics are distinct from ours. A third assumption is that these were the premises current among the Powhatans during Pocahontas's lifetime. I believe that the premises she acted from differed markedly from those that inform the modern age, which might be briefly characterized as those of a Christian-based capitalist democracy.[1]

Another major consideration in forming *Pocahontas: Medicine Woman, Spy, Entrepreneur, Diplomat* as I have is that our information about her comes from English sources. While there are a number of men who wrote about her from personal experience, their references to Pocahontas are sparse. The major source of information about her is John Smith, who mentioned her in his first report to the Virginia Company. Smith wrote a much more detailed—and probably highly colored—account of her contributions to his welfare and that of the English at James Fort, but this more detailed account was written several years after her death. By then she had been to England, been received by His Majesty King James I and Her Majesty Queen Anne, and had found much favor with them and others of the court and of English nobility. Smith, himself less than a commoner—a term that applied to those with personal wealth and its accompanying social standing—found it in his interests to portray himself as her love, her life, and her destiny. He also, like the others who reported on her, was forced, as are we all, to tell his story in the context of the understanding of his world—in his case, that of a working-class seventeenth-century English adventurer with many travels to far places and quite a bit of experience with the wealthy and noble. Smith was unique among his kind, but nonetheless was an Englishman. His accounts abide by the narrative conventions of his age and culture.

The other major source of material about Pocahontas comes from travelers who kept notes on Algonquin native languages, customs, beliefs, and material culture. Later sources, historians and ethnologists, have compiled a steadily increasing, and improving, body of material about life in the *tsenacommacah* or in other Algonquin communities of the time. Along with this cultural material, impressive work is being done by archeologists, whose work through the latter part of the twentieth century and into this one is filling numerous gaps in academic knowledge and, coincidentally, showing the veracity of Native people's accounts, from their own Oral Tradition. One thing must always be kept in mind with all such material: the objects and sites uncovered and examined are whatever they are—pottery, jewelry, foundations of buildings, and so forth; the meaning of these material remains—how they were used, what they signified to the users, how they connected with the worldview and basic assumptions of the people they arose from—can only be inferred. The interpretations depend largely, if not entirely, on the lexicon, the cultural mind-set, of the interpreters. For example, Smith could only believe that he was about to be beheaded by people he regarded as "howling salvages" uttering "fiendish shriekings," because had he been in England under similar circumstances, beheading would have indeed been imminent. Beheading was quite the fashion in Europe at the time, and the Tower of London often sported the gory heads of the executed on tall pikes, for the edification of the public at large.

In the modern era we are confronted with the exciting and exacting task of recovering America's ancestors. For the past decade, archeologists have been laboring to reconstruct the English part of America's first story at the original James Fort site. More recently, in the spring of 2003, the Werowocomoco Research Group announced the beginning of a dig in summer 2003 at a site thought to be Werowocomoco. To be conducted in conjunction with a number of

Native consultants from the original tribes of the area, this recovery is one of the most important in the history of American archeology, because Werowocomoco was the place where the newcomers were welcomed into the region by the manito. It is the sacred center of the United States.

As researchers move beyond monoculturalism into a more comprehensive interpretative stance, some things emerge: We know that Pocahontas was a Powhatan woman. We know her people were matrilocal, matrilineal, and matrifocal.[2] We know that she was a person of high status, and of unique ability. We know that the Powhatans, an alliance of at least thirty or thirty-five Algonquian-speaking tribes, were located in the tidewater region of what is now called Virginia. Their name for their homeland was *tsenacommacah*. We know that Pocahontas existed, that her name was Matoaka, that she was familiar to and with the English from childhood, converted to Christianity, wed an Englishman—a tobacco planter named John Rolfe—went to England, was received at court, and died at Gravesend, on the Thames below London, in 1617.

Some of the most interesting evidence currently surfacing shows that the Indian people of the North American Atlantic seaboard lived in houses that were constructed very much like those the English peasants and lower middle classes commonly lived in at the time: both types of dwellings were constructed of large logs filled in with small tree limbs, and each was roofed by thick branches. The Algonquin housing allowed for rolling up the sides of each dwelling in the hot summer, while the English featured thicker walls to protect against a chronically cold, damp climate.

Noticing that telling similarity opens a world of possibilities: we can see in the poetry, art, and drama of that period—the late sixteenth and early seventeenth centuries—as well as from some of the more exclusive "clubs" or communities of the time, marked similarities in thought: while Smith and other English reporters emphasized the non-European qualities of the Indians, the Indians differed little from the English of the time. In England it was commonly believed that people with certain abilities could levitate and move through the air without benefit of wings, aircraft, or broomsticks. The existence of the fabled cabal headed by Dr. John Dee and participated in by a number of important people such as Queen Elizabeth I, the Earl of Northumberland, George Percy, Thomas Hariot, Francis Bacon, Francis Drake, William Shakespeare—although that last is questionable—and Walter Ralegh[3] strongly implies that, at least at its earliest stages, from the late sixteenth century through the first fifteen years of the seventeenth, a powerful community was deeply committed to a worldview based on assumptions closely resembling those of the Algonquin: they were students of the occult and the arcane. They practiced a spiritual discipline known as Enochian magic,[4] which was based on metaphysical definitions of reality and the human place within it. While England and the rest of Europe were moving pell-mell toward the Age of Reason, when physical sciences, technologies, and modern secular states would reflect a materialist worldview, the *tsenacommacah* remained entrenched in a spirit- or manito-focused worldview.

Because these commonalities were present despite the apparent conflict between the world of Europe and that of Indian America, the effects of the latter on the former were considerable. While modern thought is only beginning to recognize that fact, more and more evidence is appearing to further support the idea that acculturation goes both ways. In this book I spend some time drawing on English traditions that parallel Powhatan traditions during Pocahontas's lifetime, because she found such a welcome in England and evinced a powerful affinity for it. It is my contention that she was influenced more by the commonalities than by the differences, although perhaps the latter delighted her as exotica delight many people of today. And it is worth remembering that one woman's exotica is another's ordinary experience.

LIVING BETWEEN TWO WORLDS

Given these various considerations, *Pocahontas: Medicine Woman, Spy, Entrepreneur, Diplomat* is a mixed-breed or hybrid study, as American Indian life in the United States is a mixed-breed or hybrid life, as I, the author, am a mixed-blood, hybrid woman. In that, this volume, like Indian Country itself, retraces the steps taken by Matoaka, a.k.a. Pocahontas, four centuries ago. Grounded in one context, many if not most Indians live lives translated into quite another. So ubiquitous is this fact of American Indian life that when translation fails, as it all too often does, a term to identify it has been coined: the "fallen between two chairs" syndrome, valorized and lamented in scores of novels, biographies, and academic studies. So common is this image of American Indian people that it is taken for granted by all those acquainted with the Native world and too often by those who reside in Indian Country.

A life of negotiating between two worlds, Indian and Anglo, is not necessarily a tragic one, though those lives that are tragic are the most interesting to write about. Millions of Native people since Pocahontas have lived such lives, whether or not they did so by choice. In the age of "realism," often the more victimized the subject, the more intriguing. Thus we are many of us very conscious of how the Native people suffered from the white domination of our world. Books such as *Bury My Heart at Wounded Knee* are classics, and rightly so. But the Native Elders remind us that ours is the period of the Seventh Generation, when reconciliation between the races is called for. In that light, the impact the Native world, in both its practices and assumptions, has had on the non-Native world is getting growing attention. From the construction of the United States as a federated democracy to the ever increasing populations of the world across the oceans from us, from the growing tendency to treat women, children, elders, and other "outcasts" with respect to the successive removal of excess clothing, we see the influence of Indian America on European thought and manners. While much has been lost, quite a bit has been gained. I believe we do great injustice to pathfinders such as Pocahontas by discounting their massive contributions to the modern world and instead considering them as having lived tragic lives, victims of European greed. While Europe was on a steep line of ever increasing desire for wealth and power that would soon engulf parts of the world so far not so inclined, there was a great deal more motivating the voyagers than income. The Spaniards were looking for the fabled seven cities of gold, which was not a monetary concept. It referred to the long leg-end that somewhere there was a land where people were healthy, where peace reigned, where immortality was a given. It was this dream that energized most of them; the money-making part was how the dreamers convinced those who had the power to decide that ex-ploring was allowed. It was a world of absolutes: absolute power of tyrannical monarchs, rapacious prelates, conscienceless hustlers, and con artists. That world was changing, as surely as was the world of Native America. Pocahontas was in a remarkable number of ways the living embodiment of this dual cultural transformation.

Such a life, like the literature emanating from it, can be intrigu-ingly complex, multidimensional in layers of meaning, and suggestive of possibilities that uniculturalism can't access. Not necessarily based on race or national origin, the life of an agent of change can happen to one born in one class and transferring to another, by being a female functioning in a male-defined world, or being an artist who must live in techno-bureaucracy. Negotiating an identity in a multidimensional world is probably the most shared feature of American life, where just about everyone is a "newbie," whether they recognize their situation or not. It is an aspect of the American character that other nations find both troubling and exciting. It is the phenomenon that has made this odd nation a powerful force for change in the modern world.

In negotiating two forms of telling Pocahontas's life, *Pocahontas: Medicine Woman, Spy, Entrepreneur, Diplomat* is designed, in terms of the organization of the five main sections, to follow more traditional narrative strategies. In the way of traditional biographies, the first major section, "*Apowa* / Dream-Vision," will center on myths and texts. Some events will be included in the story, because the story of John Smith and Pocahontas, for instance, is itself on the brink of at-taining mythic status in the American lexicon. In the Western way, though, quite a bit of the material in this chapter will be data: names, places, topography, unfamiliar terms given modern equivalents where possible, and the like.

Organizing this opening chapter has been the most difficult and challenging of the entire volume. The way the Native narrative tradition works is so alien to the way the Anglo-European narrative tradition works—formally more than popularly, by the way—that negotiating the two in an attempt to develop a reasonably comprehensible text is daunting. It is about as daunting as trying to negotiate two or more identities in a uni-identity world such as contemporary USA.

Given my druthers—that is, my instincts and sense of the story I am telling—"*Apowa*" would introduce some complex ideas that form the bedrock of Native thought, embedding historic details from the records of the time where suitable. It would include "fictional" passages, such as passages where we experience Pocahontas's life directly through storytelling, along with more abstract discussions of underlying assumptions about how the world works current among her people in her day. Her historic meeting with John Smith would be set entirely within this context—a context of tone, structure, and style as much as of "fact."

It has been said by some of the elders that until a people has clarified its place in the Great Mystery, and has discovered those mythic/ceremonial traditions that maintain its lineage, no real nationhood is possible. It is as though, they maintain, a community bereft of profound and widespread connection to the multitude of kinds of life, of intelligence, that exist all around remains a child for as long as it can survive. This idea implied that children without the benefit of the wisdom of that elder knowledge won't long survive. It is the way of all things to grow into their being in accord with what is in them and what moves through them. That which moves through a people is its heart, its manito. In those terms, the life of Pocahontas is one of the earliest documented and widely recognized indicators of what an American tradition must be, and what portion of *manito aki* is its proper medium. In that regard, I believe that it is no accident that Washington, D.C., is situated at the headwaters of the Potomac River; the Powhatans identified that river as the one most filled with manito energy, *powa,* of all the rivers north of them.

The chapters that follow "*Apowa* / Dream-Vision" grow more and more explicit, locating Pocahontas within the confines of history while maintaining a narrative line connecting the *manito aki* to the historic record. In the end the narrative focuses on the individual person, having drawn her identity as much from the contexts defining and embracing her life as from her own actions.

Throughout, whether the main focus is Pocahontas or the *manito aki,* the story is ever about the order of reality that is both beyond the mundane and that encompasses it. In order to clarify some of this material, which can so easily be perceived as a children's story, a study in primal consciousness, or a romantic version of an Indian woman's life—a kind of "Noble Red Woman" narrative—I have given some time to discussing relevant work from physics and archeo-astronomy, as well as material taken from myth, ethnography, history, and literature.

The life of a figure as great as Pocahontas, who lived in a period of great upheaval and far-reaching change in the Americas as well as overseas, is not a simple life, though for the most part it has been told as such. Nor is it a story about an ingenue, "comely if dusky," who stumbled into history by falling in love. It is a complex and multilayered story, difficult to narrate in ways that navigate the two worlds in which both her life and this volume exist. In the end it may not be Pocahontas or John Smith who are the major protagonists of the events, but a mind-altering plant native to the Americas. In the end, it seems that this humble plant was the great mover and shaker, turning the wheel of the great transformation that spanned three centuries, from the coming of the Columbus party to the end of the "Indian Wars" at the end of the nineteenth century.

TELLING A LIFE

The story of Pocahontas, as it has been enshrined in American and English consciousness, is the story of a subset: the great priestess Beloved Woman of the Powhatan *manitowinini* is portrayed as a sexy, but virginal, adjunct to Captain John Smith, the first Anglo-American hero. The *tsenacommacah,* as the tribes belonging to the Powhatan Alliance referred to themselves as well as to their home territory, comprised at least thirty-five separate southeastern Algonquin nations and is a subset of Anglo-American colonial history.

However reasonable that Algonquin point of view may be, Anglo-Americans authored the original documents and the histories and biographies drawn from them. In the worldview of record keepers and scholars, the story centers on them, the Americans of English derivation. Their version thus is not Algonquin centered, nor is it based on an Algonquin worldview. Because Pocahontas was born Algonquin and not English, and became Powhatan colonial-Virginian, not English colonial-Virginian, the full story of her life has been miscontextualized, wrenched from its own narrative ground and placed within an alien context that leaves many questions unanswered, even unaddressed. I have grounded this biography of Pocahontas in an Algonquin universe, looking to it to explain her motives as well as describe her actions, because, at the end of the day it is within an Algonquin context that we can sense the significance of her life.

A narrative tradition, like a national consciousness, has its own antecedents, common allusions, and modes of casting perception, interpretation, motivation, and action. Each narrative tradition has its cast of characters, and certain characteristics that define them. In the narrative tradition of the Indo-European peoples, a woman's adventure story is inevitably about love. Romances have been written for women and by women for several centuries, never failing to find an audience. In our day there are at least two cable television channels dedicated to them, while some of the megasales writers of popular fiction specialize in the romance. Because of this narrative convention, books and articles about Pocahontas have largely fallen into the romance genre, even when marketed as biography or history. Indeed, had it not been for those conventions we might never have heard of Pocahontas other than as a small footnote.

In the Oral Tradition, conventions apply as well. Certain themes and characters appear again and again; even the major character in a narrative might move from here to there in a seeming jumble (when seen from Western linear expectations of a narrative). A story about a given character—say, Trickster—might meander through a number of narrative cycles. In some Trickster is the main character, in others a bit character; sometimes valorous, sometimes nefarious; often senseless and therefore hilarious. Trickster can be a mocking mirror held up to listeners, or a profoundly sacred being whose name cannot be mentioned in certain seasons.

In this random, almost chaotic system of narrative, the connection of events to one anther clues the audience to the overall significance of the story being told. This may not, and usually is not, done is one sitting, like reading a bedtime story from beginning to end in one reading or seeing a movie from opening scene to final credits.

Indian storytelling sequences seem random, off the point, around the edges of what's on your mind or on the mind of someone in the audience, or on the community's mind. The references—pointed enough if you know the ins and outs, the place of joking, breaks, "illogical" shifts from one subject to another—are usually scattered about through a series of stories, conversational gambits, jokes, and odd behaviors observed among some of the local folks. All these, taken more or less together and thought about, randomly but attentively, usually over and over throughout the years, yield a surprising and continually refreshed and refreshing harvest of understanding. The ubiquitous presence of repetition—or reiteration, if you will— acts as connector and transition from one "disjoint" in the sequence to another.

Structurally—and this is reflected in the order in which events are narrated as well as in tone, stance, grammar, informality of language, formality of language, joking, and repetitive passages—the Native tradition can best be diagrammed as a series of points that can be connected in an almost endless variety of ways, depending on the story being told—that is, the points being made. Contrast this with the basic structure of Western narrative, which progresses in a chronological sequence or in a predetermined order of importance of facts, examples, and/or events. Western narratives feature one figure who dominates the story, one setting that frames it, and one point of view, explicitly stated as well as implied, that defines it. Certainty is the fundamental paradigm suggested by this structure, and this analysis is reinforced by the demand that all claims be "proven." Proof is defined as impartial witness and/or repeated replication with identical results—or citation of authoritative primary sources: witnesses and/or commentators present to testify to historical events.

However, when most of the proof is in the entire quality of the pudding, when the witnesses are as likely as not to be wind, rain, extinct forests, grasslands, long-dead buffalo, deer, clams, and the like, the proving becomes a challenge. Native scholars and writers are more and more meeting the Western traditional demands for verification of evidence or information by turning to the Oral Tradition in its narrative and other traditions (music, herbology, architecture, construction, cooking, pottery casting, etc.) to support and verify the points being made. In other words, Native scholars call on authoritative sources available to them partly by familiarity (knowing the witness, knowing her/his family for a few generations, and the like; or knowing the ceremonial tradition and understanding the deep assumptions upon which it is based).

The mixed narrative–narrator relies on a bit of both, which is how and why this book, *Pocahontas: Medicine Woman, Spy, Entrepreneur, Diplomat,* is built. Pocahontas, or so it seems to me, was one of the earliest bicultural or mixed-culture Indians of the East Coast, and one of a very few Indian women known to a general readership. As such, she represents the mixed-culture identity of the East Coast that developed out of the early seventeenth century. I decided that a book that in style as well as content reflects the mixed currents of culture and spirit that defined that era was the best approach. I hope you, dear reader, will agree.

ON THE OTHER HAND

In terms of the political exigencies of the Englishmen of the seventeenth century, contemporary narratives, the multiple *True Relations* narratives penned by various gentlemen observer-participants, were informed by and geared to the juggling for power that characterized the era. This dynamic existed not only between nations such as Spain and England; it also held sway between branches of Christianity: Roman Catholic (or Papist, as the Protestants called that venerable institution) versus Protestant; Anglican or monarch-led Christianity versus Puritanism (which saw itself as emanating from personal experience of the believer and the authority of the Scriptures; as such, Puritans espoused a congregational organization in which the elders of a town or village—that is, the property-holding "free men" of the same—chose their pastor and engaged him for as long as he reflected their beliefs and interests).

The struggle for power in what the Europeans called the New World was a game that had been defined as religious dominance rather than that of nation-state. England and Spain had been at war because the Spanish were Roman Catholic, driven by Rome, the Inquisition, and a sense of monarchy and religion being one, which meant that the pope was the greatest monarch, with national monarchs taking their power and place just below him. The English, on the other hand, had chosen the path of national religion, in which the monarch, rather than the pope, presided over the church. The Puritans were promoting a new idea, one that grew from the first: they called for a church that was independent of both monarch and ecclesiastical authority; the community of the faithful would choose its pastors; the Bible would be its authority. Until that era the dominating belief was that the monarch was the only individual and that the rest of the community—aristocrat, bourgeois, commoner, or peasant-serf—had being only as the monarch's possessions. Because of various events—such as the invention of the printing press, the decimation of populations all over Europe caused by a series of plagues, and an indefinable factor the Natives recognized as *transformation time,* a particular point in that current that comes from the world of the supernatural that surrounds and infiltrates the human world, that decrees: "Change chairs"—this millennia-old belief that the proper structure of existence was hierarchical began to give way to the proposition that no one can own human beings, neither monarch nor master, husband nor parent, church nor state. This, of course, was the primary idea that came from Indian Country to turn the Old World on its ear.[5] In the course of this narrative we will consider some of the means by which this transformation was pursued by, let us say, the Spirit of the Time.

According to hemisphere-wide Native beliefs, there are certain times when the world is changed. These periods can be determined with the use of an accurate calendar—such as the Maya Day Count—that includes the Sacred Cycles in its calculations. The realm of the sacred was/is known to the Algonquin as the manito, or *manito aki,* world of the spirit/supernatural. In English, *manito* is often translated "mystery." The Great Mystery, which is unending process, action rather than object, is gi'*tchee* (*gitchee* or *kitchee*) manito, Great or Huge Mysteriousness-interacting-always. This basic concept was soon used by the Christian newcomers to instruct the Indians in a new kind of spirituality—or so they believed Christianity to be: *Gi'tchi manito* became the Great Spirit, supposedly the Indian term for Almighty God.

In its final phase, the periods when such a cosmic transformation is afoot, a new sun is in the sky and a new earth is beneath our feet. To the Native peoples of the Americas, the period known to Europeans as the fifteenth to eighteenth centuries was a major phase in such a time. The Maya-Aztec marked it with the coming of the Spaniards to the Yucatán peninsula, as the Powhatans marked it with the third coming of the English of the Virginia Company to Chesapeake Bay. Wisdom carriers of the Native Nations tell us that the Great Transformation was completed only a few years ago, in the early 1990s. It was then the time of the Seventh Generation, the era of Reconciliation, both signaled by the birth of a white buffalo calf named Miracle.

The Changing Time, as I will usually refer to this cosmic event, saw most of Native America dissolve; it did not die. Rather, it "morphed," and is currently emerging from a long period of purging, remade. Remaking something is part and parcel of American Indian thought. And in that context the life and times of Pocahontas can be understood; not because she was the greatest of the great Indians the English encountered, but because she was an agent of change and her role in that great sacred drama was among the English.

FLUIDITY OF IDENTITY

A part of the culture of individualism, a name is considered unchanging: it identifies one from cradle to grave; only one per customer is allowed. With many Native Nations, it was a different matter, and among many it still is. The figure frozen in history bears the child name: Pocahontas. It was her familiar or informal name, but it wasn't meant to hang on as her identity; these people took (and take) a name to be indicative of one's state—and childhood is a state. When Pocahontas became a woman—menstruating, pubic-haired, married— her name was no longer Pocahontas, although those of her close family who knew her best might still use the appellation affectionately. Even among modern Americans a child's baby name is often used by the parents, siblings, grandparents, or aunts and uncles as a way of recalling familial ties, shared history, and affection.

It was as Pocahontas that John Smith knew her, and much of our knowledge about her comes from his quill. He ever referred to her as "Pocahuntis," as he spelled the name, even years after her death, and Pocahontas she remained; only the spelling changed. That modern peoples know her only as a child says a great deal about white–American Indian relations, and it reveals volumes about Anglo-European consciousness. We read our culture and believe that we are reading about something other than ourselves; it is a common enough characteristic of a race that relies on early childhood learning for understanding everything that comes along for the rest of one's life. The race I refer to here is neither Indian nor Anglo; it is human. The words *Pocahontas, Powhatan,* and *Indians* are familiar enough to many Americans. But the ideas and images they evoke differ greatly from community to community and from period to period. What *Indian* means to a person who is American Indian bears little resemblance to what it signifies in the minds of other Americans. Similarly, a narrative convention in one system often makes little sense in one that is at great variance from it. So the American Indian narrative tradition and the Indo-European one differ in a number of ways. This is a fact that certainly complicates understanding the one in the terms of the other.

That the difference is as basic as a name—name of person, place, or phenomenon—forces those intent on bridging the distance between one culture's worldview and another's to beware the easy categorization of either. *Pocahontas: Medicine Woman, Spy, Entrepreneur, Diplomat* is first and last an Indian story, requiring that readers keep in mind that the bridge we must negotiate in considering Pocahontas's life began in the Algonquin forests of the *manitowinini*. The story crosses over the sea from there, just as our hero, Matoaka, nicknamed Pocahontas and baptized Rebecca, went from *manitowinini* to Faerie, from *tsenacommacah* to England, and from child to legend.

Pocahontas, the child, is the persona who entered history when as a prepubescent girl she threw her arms around Smith and signaled that he would be adopted, or remade, into her clan. Her age at that time is reckoned at about eleven years. A few years later, she was abducted by the English and held at a rudimentary boarding school—the first of many devoted to the purpose of "civilizing Indians"—that was distant from James Fort, as the English version has it. However, from an Algonquin point of view, it is more likely that she went to them voluntarily, letting them believe whatever they would. She went as Matoaka (or Matoaks), which was her adult name. When she was adopted, was remade, to enter the Virginia Company clan, she traded her Indian name, Matoaka, for an English name, Rebecca. In Powhatan terms, in which remaking a person into another person was familiar, she was no longer Matoaka; she had become Lady Rebecca, and as Lady Rebecca she died. If we were to keep to the cultural customs of the subject, we would refer to Pocahontas as Lady Rebecca.

Lady Rebecca had another name and role or identity: Amonute. This was her medicine name, identifying her as Beloved Woman, shaman-priestess, sorcerer, adept of high degree. It was a name shared only once with the English, and as such is questioned in the pages of biographers. However, it is highly probable that she was a member of the *midéwewin,* the Medicine Lodge or Great Medicine Dance, a spiritual discipline widespread among the Algonquins all over North America. This society, or spiritual discipline, was concerned with various kinds of magic, healing being only one. The term *medicine,* like the newer word *shaman,* signifies that something Native is going on. Analogous words—that is, words that signify much of what these shamans and medicine people do, abound: in non–Native contexts such people are usually identified as seers, priestesses, priests, or wizards. Many of the actions these trained spiritual adepts take do things that defy our present understanding of how things work, do things that material laws of the old physics don't account for. Among these are teleporting objects, soul walking, rain bringing, clear seeing, prophecy, finding water or any lost object or person, even protecting soldiers from a particular community. Carrying out a World Renewal Ceremony, while not a common practice, is sufficiently widespread among Native practitioners to make it likely that it was such a ceremony, on an almost unimaginable scale, that Amonute, Pocahontas as high priestess, along with the *matchacómoco,* the Great Council of the Powhatan Alliance, was engaged in.

During the course of telling her life, I will use the name appropriate to the context, remembering, however, that most readers think she was Pocahontas, historians, biographers, lyricists, and Disney Studios having proclaimed it so. Therefore, because we are creatures of conditioning, I will use the familiar Pocahontas throughout as well.

WHAT'S IN A NAME

In regard to names, the people of the area touched by the English adventurers were, by and large, subscribers to a centralized alliance, devised for purposes of defense, and dubbed by historians the Powhatan Confederacy. However, new scholarship suggests that "confederacy" is too fixed a form to describe the rather loose system of alliances that prevailed among the southeastern Algonquins in the early seventeenth century. I have chosen several terms in this volume when referring to the thirty-five or more tribal groups involved in Pocahontas's story. One I use is *Powhatan Alliance,* implying a configuration similar to NATO of the contemporary world. Another is *tsenacommacah,* which was how the people involved referred to their allied communities. I also refer to them as the *manitowinini,* a term I borrowed from the Algonquin (Delaware Nation) historian Jack Forbes, which means "the

people of the manito." This term is particularly relevant when the
context is the subject's role in the sacred—that is, in the *manito aki*.
For the most part the Powhatans, the alliance that shifted even during
the few years before the English settlement took hold in their lands,
were *manitowinini* as much as they were *tsenacommacah*. Through it all,
and this is probably the most salient point, they were the People of the
Dream-Vision, which is what the Native word *Powhatan* means.

The story of Pocahontas is inevitably the story of the Dream-
Vision People. It is centered on a tradition immersed in Dream-Vision
protocols, lore, practices, and understandings that shaped and directed
the course of her life. *Apowa* means "I Dream-Vision," and refers to a
fairly common event among Algonquin boys and girls, as well as certain adults. It was also one with Pocahontas's role as a medicine
woman. Any biography of a Powhatan Indian woman or man from
earlier eras must include the assumption that this world we live in is
by its nature a Dream-Vision. It must assume that the manito are the
greater Dreamers, the humans the lesser, the *manito aki* the location of
the Dream-Vision as it takes shape and gains sufficient "thrust" to
move into our more physically dense reality. We can also note that
while *manito* is a cobbled-together Algonquin term for the phenomena, the awareness of such a realm is worldwide.

FAERIE: *MANITO AKI* IN ENGLISH TRADITION

In the course of understanding how Pocahontas/Rebecca found her
way into a state of high affinity with the English we need to consider
the English people's own affinity for the realm of Dream-Vision. The
English commonly refer to this realm as "Faerie" or "the Under-
world," or sometimes *Logres*. The idea implied in each of these terms
is that there is a world within and beyond the one most modern
people recognize—one that is coming to be recognized, defined, and
explored by modern physicists such as David Bohm, F. David Peat,
Werner Heisenberg, Benoit Mandlebrot, and dozens more. While
suggesting that physics/mathematics and metaphysics are closely related may cause some unease among formalists, the ancients—Greek,
Chinese, Japanese, Hindu, pre-Enlightenment European, and most

Native Nations—assumed they were closely allied. It seems to be an idea whose time has come round again. Bohm's description of the consciousness form we usually engage in as "explicate" and the other one as "implicate" corresponds neatly with the categories of material world and *manito aki,* the mortal realm and Faerie.

The state of awareness when one is in or communicating with this realm was long identified as Dream-Vision, or *powa,* in the Algonquin world. It is allied to the Native Australian concept of Dream Time, a way of organizing reality, including via sensory data, that brings phenomena into awareness that are absent from perceptual fields in another brain state. Variously known as "alternative consciousness," "shamanic consciousness," "walking in balance," "walking in beauty," "walking in a sacred manner," a state or condition of *inanyi, orenda, waken, hozho,* the "paranormal," the "transcendent"—there are as many ways of speaking about it as there are traditions in the world— this awareness state is the state entered by medicine people (who are known as Dreamers in some traditions, such as the Pomo Indians of California). Such people have a marked and proven ability to navigate the currents of the Dream; to discern, sort, identify, recognize, and act appropriately comes of a combination of inborn tendency and training. This ability goes as far beyond "feeling" or "imagining" as creating a work of art goes beyond coloring in a coloring book.

Because of the great inborn talent and lifelong training and discipline that marked Matoaka's life, I refer to her as an "adept"—one who is highly educated in Dream-Vision disciplines. Given even the scant facts we have about her, young Matoaka, nicknamed Pocahontas, must have been a shamanic prodigy, possessed of a talent and competence as great as that of the great Powhatan Wahunsenacawh. Wahunsenacawh is usually identified as Powhatan—his title, not his name—by historians and English records of the seventeenth century and also wrongly identified as Pocahontas's biological father. He was the principal priest-king of the entire Powhatan Alliance during Matoaka's lifetime. However, "Powhatan" is a title that pertains to the role as focus point or coordinator occupied by one who possessed this supreme dreaming capability.

People such as Matoaka, Wahunsenacawa, Uttamatamakin, Opechancanough, and other great figures of the time lived in a state of consciousness that is profoundly unlike the one we moderns think of as normal. I think they were almost always in an "altered" state of consciousness. Their day-to-day practices provide strong evidence of this. This is not to say they would have been identified as geniuses on modern IQ tests; nor were they psychotic. They were operating on all cylinders, using a great deal more of their brain capacity than moderns use, and thus were aware of "facts" that moderns don't consciously register.

Given that people like them were, by English accounts, of high status in their communities, it follows that they formed some kind of elite. That elite had some basis in the clan to which they belonged, some clans being of higher status than others, but for the most part it rested on just that capacity to live beyond the mundane. In a civilization based on beyond-ness as a norm, it stands to reason that those given leadership roles were the most able at learning from and acting on material they gained in that state. Central status was required for their society to run smoothly, which it did until the emerging consciousness that marks modernity landed on their shores and made its way among them.

THE IMPORTANCE OF STORY

As the Argentine writer Jorge Luis Borges tells us, "myth is at the beginning of literature, and also at its end." Or, in the wisdom of the Laguna Pueblo tradition, we are the stories. Given that I deeply believe both of these assertions to be true, I have put as many stories into this story as I could: I have borrowed from both Algonquin and English story traditions to locate my interpretation and presentation of Pocahontas's life in the story traditions that shaped it. As I said at the outset, telling a life that is neither modern nor European-American, in a language that is about as foreign as it can be to the reality the subject lived in, is a tricky operation. While there is more, much more, that can be said, and that I trust will be said, I offer *Pocahontas: Medicine Woman, Spy, Entrepreneur, Diplomat* as an honest beginning. A major rule of the Oral Tradition as I learned it is to tell the truth as best one can. Having done so, one ends one's narrative with a quaint expression that is worth pondering: "This long is my aunt's backbone."

Da'wa-e, Thank you.

Paula Gunn Allen

THE TOVVNE OF SECOTA

(printed 1590)

Engraving by Theodore De Bry

(courtesy of the John Carter Brown Library at Brown University)

Apowa / Dream-Vision

"This is not a land for gods," said the buffalo man. But it was not the buffalo man talking anymore . . . it was the fire speaking, the crackling and the burning of the flame itself that spoke to Shadow in the dark place under the earth.

"This land was brought up from the depths of the ocean by a diver," said the fire. "It was spun from its own substance by a spider. It was shat by a raven. It is the bones of a fallen father, whose bones are mountains, whose eyes are lakes."

"This is the land of dreams and fire," said the flame.

—Neil Gaiman

Mystery, not history, was the bread of life.

—Richard Noll

According to Indigenous Science, everything that exists is held within the great cycle of time, and thus there exist ceremonies that acknowledge and assist in its renewal.

—F. David Peat

oo-maa haa'a / once upon a time

That was the time that Yellow Woman was taken by Whirlwind Manito, and he took her to another place far beyond the great waters. She was standing in the south. In the south she was standing. First Whirlwind man came from the south, and then he came from the east, then he came from the east another time. That was when Yellow Woman knew that it was time. She was standing in the south, and Southwind Woman was talking to her.

Look here, Southwind Woman said. Yellow Woman. You should go to the south a little ways. You should find the one who will take you across the great water. That's the one who comes on a seabird with great white wings. The one that moves by my breath, or the breath of one of my sisters.

Yellow Woman waited. She was standing there in the south. In the south she was waiting. Then the wind came from the south, and she stood. I will not go to the water just yet, she said. Later, when the dawn comes. When the north wind comes to stay. When Ice Manito, Windigo, comes. Then maybe I will go.

Then Yellow Woman. She was very young, and she got very tired of waiting. But it was the way of her people, and her aunts kept her heart strong. They gave her many dances to learn, and many chants to sing. They helped her make the steps and the calls with her throat that would carry into the world of the manito. She went out every day. She went here and she went other places. And the time passed until the ice held the world still and silent, and she heard the wind saying. Now, Yellow Woman. This time.

She went back to where the old women were. Where the clan mothers and the aunts were. And they dressed her and put the down of white winter birds in her hair. They painted the parting of her hair red. Her face they painted the color red, pocoon, and her hands they made patterns on, and on her ankles they made other patterns.

Yellow Woman went among the elders, among all the holy people, even some of the wild men who frightened her. But she looked down or straight ahead. She remembered the chants and the steps of the dance.

They were singing. They were talking. In that Big House, quioc-cosan. *That holy House,* quioccosan. *The elders, the* matchacó-moco, *were talking. They were honoring Fire with tobacco,* apook. *Generous hands gave apook. They sang. The ones who sang that way sang. Everyone was standing. There in the Great House,* quioccosan. *We honor you, Fire Man. We honor you Fire Woman. It is truth that is spoken here. It is truth that is done here. Thank you.*

The other women. Her sisters. Red woman. Black/Dark Blue Woman. White Woman. Standing in their places. They stood in the west. They stood in the east. They stepped this way. The stepped another way. They were standing. They were singing. It was a holy song they were singing. They were crying. That the supernaturals, manito, *would hear them. That the holiest,* manitt, *would hear them. That they were standing in truth,* oo-maa'ha'a. *The voices rising.*

Yellow woman raising her voice. In the highest tremolo she was singing. She was singing. She was walking. She threw herself down. Down she threw herself. On the body of the stranger. On his body she fell. Singing in the highest tremolo.

I will take this one and guide him.

That's what I will do.

It is the truth I speak

Had Pocahontas been Laguna Pueblo and had she been a yellow corn woman and so part of the old cycle about the supernatural sisters—red corn woman, yellow corn woman, blue/black corn woman, white corn woman—this is how her story might have been told, in translation. As she was not Laguna Pueblo of the Keres Nation but Pamunkey, of the Powhatan Alliance, her story would have been told traditionally in the words and cadences of her people. Some of the words current among the people of the *tsenacommacah* I've added to this Keresan-based version. Laguna is not Werowocomoco, the town where Pocahontas was raised, and Powhatan was her language, not Keres. But then, modern Virginia bears little resemblance to the *tsenacommacah* she knew, and no one there spoke modern American English. She was of a time best thought of as world renewal

time, world change time, or world transformation time; it's clear that
this is so. The world we know bears only a marginal resemblance to
the one she knew.

HERE IN THIS SACRED PLACE I STAND

As the spring equinox approached, Pocahontas knew it was time for
her *apowa,* or Dream-Vision; it was several years before Captain John
Smith even set foot on American soil, establishing with his Virginia
Company the settlement at Jamestown. At midmorning she quietly
stepped out of the longhouse where she lived—a half-cylindrical
structure made of bark and animal skins slashed to wooden frames.
She left the central area of the village, walking out through fields
where the deer shared the path with her, across shallow streams, until
she came into a clearing.

In the clearing was a *hobbomak,* an arrangement of rocks located
in an area with a particularly powerful energy field. These fields, or
vortexes (or vortices), as they are sometimes referred to, enabled a
seeker who was properly instructed to gain information in ways that
some might call paranormal. *Hobbomaks,* many of which still exist
today, were scattered all along the Atlantic seaboard from Nova Scotia to South Carolina and were so potent that Oral Tradition cau-
tioned avoidance except under certain conditions; it was believed that
a misguided visit to a *hobbomak* could cause harm to the untrained
seeker and to the flow of the magnetic currents that held the environment in balance. Pocahontas had been directed to the proper *hobbomak,* a natural sacred structure, by her manito—her sacred medicine
power, her connection to the Great Spirits that directed her spiritual
path. She would have gone at the direction of the Council of
Women, responsible for teaching and guiding her along with the direction of her manito.

Hobbomaks contained the same *powa*—energy—as human beings,
the same energy as the deer, the trees, the fowl, the winds, and the
rivers. It is the *powa,* the intelligence-bearing energy, that holds everything in its place, within which and because of which all that is

moves into and out of an infinite series and multitudinous kinds of existences. When I was young, my mother often told me that animals, insects, and plants are to be treated with the kind of respect one customarily accords to people. "Life is a circle, and everything has its place in it," my mother would say. That's how I came to know "the sacred hoop," the term that the renowned medicine man Nicolas Black Elk used to refer to what my mother called the "life circle." What they are both getting at is the wholeness and eternity, the constant connectedness, of all that is, was, or will come to be. The circle of life reminds us that all things are of equal value. All that is equally alive, intelligent, aware, and with a purpose that is never solely individual but always inclusive of the whole. Everything on earth, from a human being to an ant, holds a place in creation that is dynamic, creative, and responsive.

The earth is alive in the same sense that human beings are alive. Not in an anthropomorphic sense. One can say with equal truth that humans are alive in the same sense that earth is alive. The concept of "sacred hoop" or "circle of life" means that the basic unit of consciousness is not the individual person or being but the all-that-moves-and-is spirit, the great intelligence of deepest space, from which all others arise and derive their vitality. Each individual being within the circle, the sacred hoop, connects with every other; and what happens to one, or what one part does, affects all within the circle. This is the way of the sacred. All that is, in whatever form it might be, is imbued with an energy of intelligence that flows in a great, ceremonial pattern. It also means that what one does must always be because everything else did/does too.

The life circle is composed of both space and time; everything is moving, interacting, communicating, exchanging information/energy. In the traditional understanding of most American Indian systems, including the Algonquin, the traditional concept of time is timelessness, as the concept of space is multidimensionality. In the ceremonial world—the *manito aki,* or spirit world—space and time constitute the field and matrix of All-that-is, including human society, and within it the circle of an individual human's life. The Grand Medicine tradition of the Algonquin, *midéwewin,* is a method for directly experiencing this otherwise very abstract concept; the institution of Dream-Vision is another.

The English Anglo-Celtic universe that Pocahontas entered that day in 1608 when she threw her slight, child body over John Smith to designate him a candidate for rebirthing, differed little from her own. The idea that the realm of the spirits or supernaturals was powerfully engaged in the day-to-day life of nations as well as of villagers was commonly held on both sides of the Atlantic. For both, arcane sciences were based as much on astronomy as on morality or piety. Both predicted events by the position of certain stars on the ecliptic plane around earth as much as by visionary techniques, and both assumed the reality of malicious as well as beneficent supernaturals, or spirits (manito to the Powhatan).

The Salem, Massachusetts, witchcraft hysteria, which led to several deaths at the stake or by drowning, and which occurred in the last years of the seventeenth century, bear witness to the strength of Puritan convictions, which mirrored the convictions held by the Algonquin. In England and across Europe, there were physical locations such as cathedrals, fairy circles, standing stones (menhirs), and springs that were seen as receptacles of an energy that could alter consciousness, at least temporarily; London itself was rife with such locations. Similarly, such places were in use among all the Algonquin; some of these were naturally occurring, such as the Powhatan River—currently known as the James River in Virginia—and the Potomac, which runs through the capital city of the United States, Washington D.C. The Powhatan considered the Potomac, which formed the northern border of *tsenacommacah,* the most power-filled or mind-altering of its rivers. *Hobbomaks* are found scattered all over Algonquia.

The *hobbomak* provided Pocahontas the ideal setting for her Dream-Vision. Just as the identity of Pocahontas as a Beloved Woman was known before her birth because of a Dream-Vision of one of the elders, and shared among the *sunksquaa* and *sunks* (wise women and wise men), Pocahontas's Dream-Vision would inform her of her place within great events and changes ahead. Dream-Visions are maps for navigating one's life path. Composed of messages coded in the language of the *manito aki,* the spirit world, Dream-Visions came accompanied by a guidebook, the Oral Tradition; and by travel guides: one's guardian spirit, or *powagan,* as well as one's *midéwewin* teachers/sponsors. As the pathway to the spirit world, such dreams cross the boundaries of ordinary time and offer answers to the myriad problems that face the nation, clan, or individual. Dream-Vision rituals involving deprivation of food and drink could pull the dream forward to the dreamer. The Powhatans held that the events or occurrences in a Dream-Vision must be enacted, so that the manito could enter the dreamer's life and reshape it in accordance with the laws of harmony, or, as the Navajo people refer to the concept, *hozho,* the way of beauty.

She was trained from early childhood in the sacred ways of a Beloved Woman—a certain kind of medicine woman or priestess—because her birth name, Matoaka or Matoaks, is thought to mean "white (or snow) feather." Since a white feather, or numerous white feathers, always signifies a Beloved Woman and is carried or worn by such women most of the time, it is likely that she did indeed have that "calling," or vocation, from birth. Her clan name, Matoaka, signified her station in life—her destiny, if you will. It foreshadowed the part she would play in the transformation of the Powhatan peoples and of the land they knew as the *tsenacommacah.*

From childhood she also would have learned the ways of the *midéwewin,* the Medicine Dance as it is called in English, as well as other religious practices of her Powhatan people in addition to her training for her duties as Beloved Woman. One of these was the Dream-Vision. Commonly among the Algonquin everywhere, young people of her status were usually given, or gifted with, a vision. It came to them as a gift from the manito—the powers, beings, forces, reality, if you will, of a world that is not quite this one, but is bigger, beyond, beneath, before, behind, and above this one. Those who gained a vision were thought to have matured sufficiently to "walk in a sacred manner," as the Lakota put it; to "walk in balance," as the Pueblo people of the Southwest say. To know how to act respectfully in the world.

For her vision, if it was the custom of her people, perhaps she left her hair purposely uncombed and dirty; perhaps her face was smeared with ashes. Her chest and back and face would have been painted in certain patterns, the colors relating to her personal journey as well as including those always worn by young women; her wrists and ankles would have been tattooed with the distinctive patterns of her Pamunkey clan (the Pamunkey were among the most powerful tribes within the Powhatan Alliance). Her nickname or child name, known informally to all, was Pocahontas. It is the name by which we know her today. While its meaning has remained unclear, it is related to a kind of vivacity, mischievousness, and quick intelligence. The name, at least as it was understood by the English, may have been related to the chipmunk or rabbit, both animals recognized as tricksters among her people. The attributes implied in her nickname may have alluded to her *powagan,* her spirit guide, for they would have certain characteristics in common which is why she got that nickname. She might have been bidden by her animal relative and guardian guide to the place of her Dream-Vision. The chipmunk was seen as a quick thinker, intelligent and resourceful, known for her proclivity for taking chances, and the rabbit was mischievous, but also creative and shrewd, possessing a wild sense of humor. *Pocahontas* was translated by the English in a number of ways, all of which reflect the playful, spunky characteristics of the chipmunk and rabbit: "wanton," "mischievous," "sportive," "frolicsome," "frisky." These animal markings would have reflected her power. Drawn from her manito, her power or guide was a kind of personal friend and mentor from "the other side." The spirit would help her, guiding her through her ordained tasks.

Dressed and adorned to show the great powers her respect for them and her poverty before them, Pocahontas might have entered a *hobbomak*. The configuration for a personal vision would have been small, most likely resembling a small hut, although *hobbomaks* of many designs have been studied. There Pocahontas would stay for a certain length of time, probably between four and seven days. She would have spent the time alone—that is, without human company, although her human guide-teacher would have come near every twelve or so hours to check on her. She would remain in the place of power, a magnetically charged sphere that the *hobbomak* generates, for a specified time, or until she was gifted with a vision. Within the *hobbomak* or other vision-dreaming place, she would have honored the spirits and entreated them to send her a vision. It might show her something significant about the path she was destined to walk or it might be more concerned with the community as a whole; it might be straightforward and its meaning obvious, or it might be obscure, filled with symbols and twists and turns that left her disoriented. It might have been very funny, or sad, or evoked a variety of emotions. When she was done she would tell it to the elders, the teacher or teachers directing her training, and they would together work out the meaning of her vision dream.

In her Dream-Vision, Pocahontas found herself by the water. She saw several ships, like great white birds, coming into Chesapeake Bay, with strange people pouring out of them. She also saw wigwams, fields, medicine houses, the towns of Pamunkey, Mattaponi, Powhatan, Potomac—all that was familiar and dear to her—shrinking, made as tiny as ant villages, even sinking under the hills and riverbanks. And she saw her people disappearing.

She came upon great piles of shells heaped onshore all around the bay. In her vision, she moved closer and closer to the piles until she discovered that the heaps were not shells at all but bones. She looked down and in her hand lay a perfect quahog shell, its outer rim purple, its interior glossy smooth and gleaming like a purple pearl. As she watched the shell upon her open palm, the shell turned into a white feathered fan, three feathers from a swan, white as the wings of the sails that kept coming into the Tidewater shores.

Pocahontas then saw a strange man float up onto shore, his body lifting and falling gently on the quiet tide until he came to rest on two low flat stones. His head was turned toward her and she looked directly into his bearded face.

Her Dream-Vision ended. As Pocahontas returned to her day-to-day presence she realized that in one hand she held a perfect large wampum shell, a white mollusk with a purple edging and pearly interior. Her quest was successful; the manito had spoken with her.

While there is ample evidence that Dream-Visioning was a frequent practice among many of those nations grouped as Algonquin, as well as among many other Native nations, the specific vision that may have come to Pocahontas must remain unknown at least to the extent that neither she nor a witness such as her human guide reported it. Nevertheless, its likely contents can be reconstructed from the events of her life, those that involved her personally and those that her people—both native and adopted—experienced.

THE ERA OF TRANSFORMATION

There had been prophecy known among the *tsenacommacah* since, it was said, ancient times. As the English writer and adventurer William Strachey reported in 1610 during a visit he made to Jamestown:

> There be at this tyme certayne Prophesies afoote amongst the people enhabiting about us . . . which his [Powhatan's] priests continually put him in feare of. . . . How that from the Cheasapeak Bay a Nation should arise, which should dissolve and give end to his Empier.

Strachey also reported a second prophecy that held

that twice they should give overthrowe and dishearten the Attempters, and such Straungers as should envade their Territorye, or laboure to settell a plantation amongst them, but the third tyme they themselves should fall into their Subjection and under their Conquest.[1]

In other words, this ancient prophecy was a heads-up: the days of the Powhatan Alliance were numbered. It let them know that even though the first two attempts would be foiled by the Native residents, still the "envaders" would triumph. For the English, the "third time would be the charm," like in an English folktale. Had they been aware of the aliens' profound belief in the magical properties of the number three, that alone might have given the Algonquins a hint: these would not be of any Native Nations, but invaders so foreign that even their magic numbers were not the same. For among most of the inhabitants of the Americas, four is the sacred number, and so, among them, "the fourth time is the charm."

As the prophecy warned, the third invasion would result in the subjection of the Algonquins of the Chesapeake Bay region and the dissolution of their "empyre." It is puzzling that a prophet, ancient to the Powhatans of the sixteenth century, could have foreseen a new nation arising.

Things were definitely changing in Powhatan country. The change facing the allied tribes was massive, and they were sadly aware of the fact. They knew it was the time of the Great Renewal because they had a certain way of telling time that depended on their traditional ways of knowing, a body of carefully preserved astronomical information, and the procedures to use to determine major events.

The allied communities of the Powhatan Alliance shared with other Native American peoples a particular gift: the ability to prophecy with remarkable accuracy. Such was the setting into which Christopher Columbus sailed on his fated transatlantic journey in the late fifteenth century. By the time Cortés marched his scant force of soldiers and burden bearers a few years later into the great city of Tenochtitlán, Montezuma II, supreme wise man and speaker of the Aztec nation, was well aware that their civilization was at an end. Didn't the sacred mirror reveal the great Quetzal, the green-feathered sacred bird, spirit form of their ancient prophet Quetzalcoatl, flying away from the sacred city? It meant that their end time, told in the stars and foretold in their traditions, had come. What would look like their doom was written upon the stones of their temples and in the friezes of their public buildings.

Columbus didn't cause the change. He came across the water to fulfill prophecy, although neither he nor most of those who would follow in his wake would realize it. The prophecy was centuries old by the time of Columbus, and when the Great Priest Montezuma heard the voice of the Serpent Woman, great Cihuacoatl, the Mother of Life and Death, crying loud in the streets in the hours before dawn, "Oh, my beloved children, where can I hide you?" He knew the time of the Great Change had indeed come, just as the court astronomer-priests had informed him. The change was in the stars, as well as carved on the sarcophagus of the sacred Quetzalcoatl; the doom of the Native American world's great civilizations was spelled out in the tangible proofs that devastated Montezuma, and made him ready to yield—not to Cortés and his small-statured, grubby handful of Spanish soldiers, but to the time.

The Aztecs were a people of many gods. Chief among them were two female deities: Cihuacoatl, Serpent Woman; and Coatlique, Serpent Skirt. A god named Huitzilopochtli, also known as Smoking Mirror, was the son of Coatlique, and it was to him that humans gave their lives in a dreadful ceremony. There were two or three ways in which those who would offered their lifeblood to Huitzilopochtli. The major one was by being personally chosen by Serpent Skirt.

One might be sitting in the square, sunning oneself, or displaying food, golden wares, pottery, cloth, or similar items for sale. At some point one would discover a flint knife blade concealed in one's garments. No one was seen to place it there; it just appeared. This knife blade was a gift from Serpent Skirt, and it meant that someone in the recipient's family, perhaps they themselves, perhaps another, had been chosen.

The importance of this ceremonial offering to the story of Pocahontas is that even among a people who history has depicted as male dominated, the power of life and death fell not to the priests, not to the soldiers, not to the courts, but to the great Feminine Principle, the living intelligence we moderns know as magnetic fields or "telluric currents."

THE REMAKING OF JOHN SMITH

On a December day in 1607, the great Native American cosmic clock, the movement of the stars, and Pocahontas's Dream-Vision all pointed toward a moment of decision. And on that day Captain John Smith, newly elected to the board of directors of the Virginia Company, set out to complete some unfinished business. Smith, born circa 1580, was twenty-six years old when the first expedition landed at Jamestown. A farmer's son, Smith had already led an adventurous life before arriving in Virginia. He had fought with the Dutch army against the Spanish and in Eastern Europe against the Ottoman Turks, when he was taken captive and enslaved. He later escaped to Russia before returning to England. Smith was a new breed of Englishman— thrill seeker, entrepreneur, man on the make. While Christian in faith, he was more interested in adventure than spirituality of any sort. He was, for his time, a kind of "new man."

Serving as president of the Jamestown colony from 1608 to 1609, Smith required the colonists to work and trade with the Indians for food. The Virginia Company, suffering from disease and starvation, had dwindled alarmingly. Weakened, they had yet to complete two major goals: to locate the headwaters of the Chickahominy River and to contact Powhatan, the "king" of the Powhatan Alliance. An expedition was organized to accomplish their tasks. Smith was awarded leadership of this perilous mission by chance, having entered a game of dice.

Choosing to go upriver by barge because its low draft made it more navigable in the shallow waters of the Chickahominy River,

Smith and some others set out, alternately sailing and rowing north of where the Virginia Company had set up camp. After some miles the river became difficult to navigate. Smith decided to continue the journey by canoe. He selected two of his men to accompany him, one a carpenter, the other a gentleman. With two Chickahominy men as guides, he set forth again, leaving the remainder of his crew on the barge securely at anchor in a wide bay. The Chickahominy were an ally of the powerful Chief Powhatan, and though the tribes shared many similarities in culture the Chickahominy remained independent of Powhatan's direct rule. They were one of the first tribes to establish trade with the English, exchanging corn, meat, and other necessities for hatchets, beads, and trinkets. Smith communicated to the Chickahominy men he had hired that he was going "fowling," hoping to keep his true intention hidden.

Traveling upstream in a canoe large enough for several men, the men continued on as the river narrowed and eventually became obstructed with fallen trees, brush, and other impediments to further navigation. Smith decided to proceed afoot, leaving the two Englishmen and one of the guides with the canoe. He told his men to keep their matchlocks ready—and to fire a warning shot should trouble develop. A few minutes after he and his companion had moved out of sight of the canoe he heard the loud cry of attacking Indians but no warning shot. A large contingent of Pamunkey men had set upon the Englishmen at the canoe, killing them.

Realizing his danger, Smith grasped his guide and lashed him to his own wrist, using him as a human shield. Keeping his other hand free to use his pistol, he turned to face his attackers, but realizing he was up against great odds, he began backing away, pulling his hapless guide between his own body and the flying arrows. So intent on his backward progress was he that he didn't realize he was entering a bog, in which he lost his footing and went down, dragging the Chickahominy man with him. Finding himself with no recourse, he surrendered his weapon to the Pamunkey war party and was taken into custody.

The Pamunkey captured Smith, believing that he had been dropped fortuitously into their hands, which was indeed the case. Perhaps because of Pocahontas's Dream-Vision years before, perhaps because the manito had more recently notified the *matchacómoco,* the Grand Council, of the right candidate's approach, they knew they were at a turning point. The coming of the Virginia Company to their lands marked the third incursion of strangers, as the ancient prophecy had warned. And though Smith would insist that his company was in the neighborhood only temporarily while they evaded Spanish warships, the Indians were aware of the potentially dreadful consequences of that third landing. Given their strategy for dealing with this ultimate crisis, one must conclude that they agreed that this ending would be the time of renewal, of the *tsenacommacah* and of the *manito aki,* of Turtle Island and whatever stream of reality it swam in.

The Powhatan prophecy had foretold of three invasions and Smith and the Virginia Company were the third and biggest threat. For the Powhatans and their allies had already wiped out the first two: Roanoke's "Lost Colony" in 1587 and another botched attempt north of Chesapeake in approximately 1595, about the time Pocahontas was born. That the stranger had fallen into the hands of the Pamunkey men who were of Pocahontas's tribe speaks volumes, given the events that followed in the weeks after his capture. There may have been some thought among the *matchacómoco,* the Grand Council, that one of the budding Beloved Women they were training at the ceremonial center of the *tsenacommacah,* a town named Werowocomoco, might be the one the manito would move to "remake" him— that is, to "give him birth" as one of the *tsenacommacah,* no longer stranger and threat, but relative, bound in law and loyalty to the tradition that held them all within one common being.

Among the powers invested in a Beloved Woman was the authority to decide who among captives would live and be adopted into the tribe, or die. The ones so chosen weren't being "saved" from death so much as dying in one identity and being transformed into another. For Smith, this would mean that he would be subject to the societal laws and traditions of the Powhatan people, and the small Virginia Company would be spared because they would become a subsidiary tribe or semiautonomous body within the great Powhatan Alliance.

Smith's capture came just prior to the time of the great ceremony Nikomis, named for the Woman Who Fell from the Sky. At the ceremony, this stranger could be brought before the leading members of all the tribes of the Powhatan Alliance. If Pocahontas recognized him, he would be remade as a member of the tribe. But if he was not the man in Pocahontas's Dream-Vision, there would be immediate war, and their dream of survival, of the continuance and renewal of the *midewéwin,* the Grand Medicine Way, would be dashed. And Smith would be taken out of the *quioccosan,* the Great House, and killed.

The Pamunkey men brought Smith to Opechancanough, one of Chief Powhatan's brothers, who then escorted the captive around Powhatan country. Smith was taken from one village to the next, following an irregular route. At each stop he was entertained and well fed, often with sumptuous feasts. These feasts were in fact part of the necessary ritual preparations for the great event to come, though Smith of course had no way of knowing this. After he left each tribal village, the *weroances* (male leaders) and *weroanskaas* (female leaders) met and concurred: This man and his people were the threat depicted in Pocahontas's Dream-Vision.

Smith offered his captors glimpses of the equipment he commanded. Though he was not particularly versed in the body language of the Powhatan people and largely ignorant of their spoken language, the Indians' awed response to his compass ("excitedly they repeated 'manit, mannito'") made him feel smugly superior. This buoyed his flagging spirits, because he was in great peril, as he saw it. As far as he knew, all the men with whom he had traveled upriver were dead. The handful remaining at the Virginia Company settlement were in no condition to search for him or extricate him from his captors. He believed that he had been betrayed by the Native guides, and without their help, he knew, any settlers who tried to find him would never succeed.

After six weeks of his tour of the villages of many of the tribes that formed the *tsenacommacah,* John Smith finally achieved his main objective: meeting the man he knew as Powhatan—a task that he had been ordered to do months before by the governor of the James Company at James Fort. By the time he entered Werowocomoco, the large town where the *tsenacommacah* maintained their central governmental seat, his selection—for execution or adoption—had been signaled throughout the alliance. There was great anticipation among the people. Elaborate preparations had already begun for the Great Feast of Nikomis, to take place over the course of a week, including the night of January 8, when Smith would be brought before the people. The leading members of the thirty or so nations of the Powhatan Alliance, who designated their greater community and the land it encompassed as *tsenacommacah,* would be packed into the generously proportioned *quioccosan,* Great House, the Powhatan Alliance's most important ceremonial structure. This building—or temple, as the English identified it—was of awe-inspiring proportions. At over two hundred feet in length and over fifty feet wide, it dwarfed many an edifice in England. At one end a huge carved statue of Oke dominated the area where the major ceremony would take place. Along each wall were two tiers of benches upon which would sit the hundreds of dignitaries participating in the ritual. In the center was a huge fire, into which all tossed handfuls of tobacco, filling the room with the odor that the manito so loved, ensuring their holy presence during the proceedings. Four great carved pillars supported the roof, each painted black on the west and red on the east. At the top of each the awesome face of a manito looked out upon the proceedings, the visage that had already been present in the chosen tree trunk enhanced by skilled carvers whose personal *powa* and training made them particularly fit for so sacred—that is, dangerous—a task.

NIKOMIS AND THE SACRED CREATION

A sacred enactment of Sky Woman's fall to Turtle Island and the creation of the world that accompanied her fall, Nikomis was a major ceremony. Like many ceremonies, Nikomis—Grandmother-dance, you might translate it—maintained the continuing connection between human beings and the spiritual world, but it also provided the *powa,* manito energy, for world renewal or transformation—when transformation was the order of that period. This winter festival would be more significant than many in the past. The people had come to the end of their ways, at least their ways over the past cycle. What was to come was being settled now, in the *quioccosan,* Great Medicine House. Pocahontas might have thought that she was an intrinsic part of this unwinding because of her Dream-Vision. But whether or not the ceremony called her out as the new man's sponsor, she would dance tonight and for several days and nights to come. Then the new way would begin its journey into the light; when it was time ripe, its form would be revealed for all those of the Wisdom Path, the *midéwewin,* to see.

Ceremonies such as that that marked Nikomis, the midwinter feast, were designed to manipulate or direct spiritual energies toward some larger goal, to transform someone or something from one condition or state to another. Thus, a healing ritual changes a person from an isolated (diseased) state to one of connectedness (health); a solstice ritual turns the sun's path from a northerly direction to a southerly one or vice versa; a hunting ritual turns the hunted animal's thoughts away from the individual consciousness of physical life to total immersion in collective consciousness. The ritual allowed the sacred transformation to take place so the traditional balance of dying, birth, growth, ripening, dying, and rebirth could continue. In the transformation from one state to another, the prior state or condition must cease to exist. It must "die." During the Nikomis festival, John Smith would undergo just such a ritual—magical—transformation. The purpose of this "adoption" or remaking ceremony is to magically change an Englishman into a Powhatan. If the magic works, the new man will belong to the *tsenacommacah.*

FEAST OF NIKOMIS /
WORLD RENEWAL CEREMONY

The *weroanskaas* prepare Pocahontas and the other youngsters for the ceremony. They braid tens of swansdown feathers into several girls' long hair, while several others' hair remains unadorned. These girls are something like novice priestesses; they are learning the ways of their professions, which for some will be Beloved Woman, while others might be medicine women or the head of their respective clans. Clans are not tribes; they are kinship units, often connected to some vegetable or animal-spirit *powa,* sacred energy, and clans may be divided among several villages or tribes. That is, Oak Clan might be common to several tribes such as Pamunkey, Chickahominy, and Mattaponi, for example, and in each community there might be a clan head, who is always a woman who inherits the position through her maternal line and because she has a certain kind of character or potential. Members of a clan, whose membership is also through the maternal line, never marry or have sex with members of the same clan; nor do they marry members of clans who are affiliated in a certain way with their clan. Anthropologists have identified this multi-clan grouping as "moiety," and many tribes have two moieties, winter and summer. Others are divided in different ways, but however the boundaries are determined, they have marriage laws associated with them.[2]

The feathers she wore in her hair signaled that she was of sacred lineage by birth and by training, and had the sacred gifts bestowed on some women by Sky Woman, Nikomis, whom the great feast honors. The Powhatans trace their origins to the descent of Sky Woman to Turtle Island (the earth). The sacred narrative of Sky Woman tells the story of how white swan feathers became the symbol of the Beloved Woman and explains how the advent of Sky Woman is an indicator of times of great change. As the designated representative of Sky Woman, Pocahontas, like her stellar ancestress, would give birth to a new world.

The world Pocahontas lived in began with the descent of Full Bloom, Sky Woman, to Turtle's back. This event, rendered variously as a mythic narrative and a sacred ceremony, is most accurately understood as an astronomical event. It was a primary case of "standing in a sacred place"—which is the same as "dancing in a sacred time." It was one of those world-changing times that might be separated by one or two thousand years or more. Such times are caused by an event recorded in the stars. The "sacred" energy of such events is held in certain stories (called "myths"[3] in English) as well as in periodically repeated ceremonials.

As it is recounted in one of its myriad permutations, a woman of "great medicine" (a term that probably is a corruption of the original, *midéwewin*) power had spoken with her dead father. He had told her to marry the *weroance* whose people lived downstream. Before he would accept her as his wife, the *weroance* put her to a series of cruel tests that, in the end, proved her medicine greater than his. Rather than returning with her new husband to her own people, because his people's customs were different from hers, she went to live with his. Soon after she settled in, the *weroance* took sick. His advisers claimed that his illness was due to the malevolence of his new wife and counseled him to rid their town of her. Doing this proved difficult because her powers were greater than his. After a time his counselors came up with a plan.

The *weroance* was renowned for a beautiful tree of many lights that grew just outside his home. Its blossoms shone brighter than the sun at midday, and it was from this tree that he derived his great power. It was hung with the flesh of his previous wife, who, though a powerful *weroanskaa,* proved weaker than he. Still, by virtue of the dead woman's medicine, which he had captured and held contained within the aura of the great tree, the blossoms gave light more beautiful than the Milky Way at night, more radiant than the sun during the day.

The advisers instructed him to uproot the tree of light, then call his new wife over to see what had occurred. When she did so, he was to persuade her to bend well over the pit that its uprooting would reveal, and when she did so, he was to give her a great heave, knocking her into the abyss. Grabbing whatever she could as she began her fall, she clutched only bits of corn plant, tobacco, and some root tendrils caked with the rich soil the tree had been embedded in.

She fell into the void, falling forever, it seemed. She fell through the darkness. There were some waterfowl floating and they saw her plummeting toward them. As one, they closed ranks, forming a nest of their spread wings to catch her in. When her fall was thus broken, they began to wonder what to do. Their conversation was overheard by Great Turtle, who came near and offered her back as a place where the fallen woman could safely stay. Because of the flurry of activity, other water dwellers came. They offered to dive into the depths and retrieve the bits of mud that had fallen from her hands and hair as she fell. Otter tried, then toad. At last muskrat dived so deep that when he finally emerged near Great Turtle he was dead. But he had stuffed his mouth and filled his paws with soil. This they took and patted around Sky Woman, and when she awakened it had spread out all around, as far as she could see.

Full Bloom had been pregnant when she was so precipitously evicted from the Sky World, and in time she gave birth to a daughter, Winona, or Blossom. In time, Winona also conceived, perhaps by a wandering manito, and in her time gave birth to twin boys. The elder, Ahone, emerged onto Turtle Island in the traditional manner. His brother, usually called Oke (Okee, Oki), took a route he found quicker and more in keeping with his idea of propriety: he tore his way out through her side, killing her in the process.

In the original *apowa* remembered as the Woman Who Fell from the Sky or the Sky Woman Myth, which happened "time out of mind," as the elders say it, Grandmother, Nikomis, removed her daughter's body, taking it to her own house, hanging it from a great tree that flourished there. Soon the head of her dead daughter became the moon, while her body became the sun and stars. At each ritual reenactment something similar occurred. The institution of a new tree of light in a new place marked the end of a particularly sacred cycle, one of Creation within the between, occurring in the *manito aki* and bringing its power into action to create what would, long after that phase had been completed, be the Turtle Island that humans inhabited.

In the midwinter ceremony in 1608, the child Pocahontas became Winona. This meant that in sacred time her head would become the new moon, and her body the new sun and stars. So the cycle has ever continued, and the way of the Sky World, the *manito aki* or Other World, is always renewed. It could be said that the Sky World is far from the Turtle Island, because understood one way that is true; understood from another perspective, of course, the faraway place is Dream Time, the place the initiated enter during *apowa* or *powwaw*—a single person's Dream-Vision or a Dream-Vision shared by a community of initiates—and it is as close as the sky.

Like most if not all ceremonies—or "dances," as many Indians call these events—the fall of Sky Woman is linked to astronomical events and constellations, in particular the constellation designated "Swimming Ducks," or waterbirds, identified by the Skidi Pawnee. (The Western constellation Scorpio contains these same two stars, called Lambda and Upsilon Scorpio, which form its stinger.) The Beloved Woman is always marked by the presence, in her hair or hands, of swan feathers, as a symbol of her connection to the waterbirds constellation. One of the tenets of the Powhatan people signals that when an event heralded by the coming of Sky Woman, Nikomis, to Turtle Island recurs, as it does on a regular basis, sometimes thousands of years apart, the world is on the brink of massive change or renewal. Given the events that began that night in January 1608 C.E., the configurations of the stars and planets must have been in the same relative positions as they had been the last time such a great ceremony was held. Not that Nikomis wasn't celebrated every winter; but this particular January the stars would indeed fall, and everything would change.

The Great Feast of Nikomis begins. Two to three hundred people, all leading members of the thirty nations of the Powhatan Alliance, are invited—priests, *weroances, weroanskaas, sunks, sunskaas*—and all have been fasting for several days. The ceremony takes place in the Great House, a long structure fifty to sixty feet long and shaped like a Quonset hut, a half cylinder made of wood, with an arched roof covered in huge birch-bark shingles. There are two doors, covered with flaps of woven reeds, one at each end: the east end is for entry; the far end, the west, is for exit. Two great posts, painted red on one side and black on the other, hold up each door. Those attending the ceremony will enter from the east, so that the red is facing them. At the center of the great hall is a great fire. There is a small hole at the top of the roof to let smoke out, with a flue that can be adjusted for the wind. Four more great posts hold up the roof and are spread at least fifteen feet apart, flanking the fire in a great square. They are also painted red and black. Spirits live in these posts; the Indians have carved away the excess wood to reveal the faces of these east and west spirits.

The great Powhatan and the Queen of Appotamac, a member tribe of the Powhatan Alliance, enter first, as a couple. Powhatan sits halfway down on the north side of the hall, in the center of the first of two rows of long wide benches, five or six inches above the ground and wide enough to sleep on. The queen sits across from him, also on a bench that runs the length of the Great House. Two by two, pairs of men and women—the *weroances* and the *weroanskaas*—enter with a ceremonial Indian dance step. The men, on either side of and behind Powhatan, fill in the two rows of wide benches, along the north and south sides of the Great House. The women sit on a second tier of higher, slightly narrower benches. They wear headdresses and crown-lets of white clamshells, which signify that the ceremony is a time of rebirth, as well as pendants, cloaks, mantles, and chains of pearls; these adornments are badges of distinction worn on special occasions. The *weroances* also wear large plates of copper, considered a precious metal at that time, attached to their heads with sinew and string woven from cotton. On the way to their seats, the elders stop to drop a pinch of tobacco into the fire to honor it.

Following these pairs, the male warriors file in, again representing the thirty tribes of the Powhatan Alliance. Then two groups of twenty girls each enter; one group sits on the west side by the exit door, the other on the east side by the entry door. Pocahontas is in this second group. These girls, Beloved Women and Beloved Women in training, are wearing swansdown and feathers in their hair. Their foreheads and heads are shaved on both sides, with one long plait of hair that hangs down to their waist or below. All the women present are painted and tattooed with floral designs. They wear very fine deerskins adorned with beaded wampum shells or floor-length capes made of turkey feathers. The men and women all have painted their heads and shoulders with a red pigment made from puccoon root, dried and then crushed to a powder and mixed with walnut oil or bear grease, signifying rebirth/sacred earth/hope/renewal.

Among the several priests in attendance is an officiating priest, whose skull is shaved except for a ridge of hair on the crown of his head, like a rooster. Bangs around the front of his forehead have been stiffened with bear grease so they stand straight out like the bill of a cap. A topknot of hair at the very crown of his head is arranged in a kind of nest of stuffed snakeskins and the skins of weasels and other animals, tied together by their tails, which meet at the top in a crownlet of feathers. His body is painted half red and half black, in keeping with the solemnity of the occasion. He wears a garment best described as a full petticoat that extends from neck to midthigh, with the waistband around the neck. The garment evidently has openings for the arms.

The priest addresses the fire in prayer: "You promised that our way would go on forever and we ask you to keep that promise." Then he addresses other powers: Grizzly Bear, Turtle, Snake, Eel, Rattlesnake, Night Spirit. "You have blessed us many times when we fast." He then prays to the Great Renewer, Nikomis, Grandmother, and his final prayer is to the Chief of the Manito, saying, "You who blessed us and said you would help us, I offer you tobacco and ask you to let our people live. And if we are about to be destroyed, I ask you to prevent it." As the priest says his blessings, everyone present shakes gourd rattles and they sing and dance. Some have turtle-shell rattles around their ankles. Songs, prayers, music, dances, ritual movements, and dramatic address combine to form an integral whole in the ceremony.

The atmosphere of the Nikomis ceremony grows intense. The moment has come for the stranger to be brought into the Great House. His escort is Opechancanough, who had escorted Smith through the villages on his six-week walkabout. Opechancanough brings Smith first before the queen, who offers him food, thus blessing and honoring him. The Indians always offer food to a newcomer, a gift exchange that says, "I accept you." One of the *weroanskaa* approaches him and washes his hands. Another dries his hands with feathers. Then a long silence falls upon the Great House. The rattling of the gourds is stilled, the great fire glows, and all that can be heard is a low murmuring among Powhatan and the elders seated with him, engaged in a kind of call and response. The air is thick with heat and ash. Powhatan suddenly stands up and raises both his arms; he commands that two great flat stones be placed directly in front of him. Two men bring the two stones, which are sacred and have been blessed. One is painted red, the other black, signifying a transformative state between two worlds. Those in attendance at the ceremony do not know what is going to happen. The men step back into their places.

In a moment, Smith is surrounded by a group of men— Powhatan men are quite a bit taller than most Englishmen—who grab him up and drag him toward the Powhatan's feet. There, they lay his head upon some flat stones. One of the priests has a club, made of roots, carved and painted so that the spirits within can be seen. Another has a club made from a straight stick like a dowel with a small plow "blade" on the end that looks like carved stone but is actually clamshell. According to tradition, whenever a new soul was formed, a particular gesture made by the official with a clamshell completed the transfer; new life begins.

To Smith, however, his prone position on the ground, his head against flat stones, can mean only one thing: beheading. There stands the man Smith thinks to be the executioner, club raised threateningly over his shoulder, his posture and gaze directed at Smith's neck, looking as if he will beat his head to a pulp. Smith himself had witnessed with his own eyes hand-to-hand combat in which warriors would catch their enemies by the hair and beat their brains out with clubs.

The priest now gestures toward Smith—fiercely, as Smith recounts. It seems to the helpless Englishman that he will not be spared.

At this precise moment, as he tells us in his account, a high-sounding wail pierces the air. The voice, from the far end of the Great House, is Pocahontas's. He hears it as a hellish shriek, but it is probably tremolo and very high, perhaps well above high C: intoning that is given measure by moving the tongue against the palate to a beat of about a second or two in duration. In Indian country it is a familiar and welcome sound. Evidently that was not the case in the England of John Smith's experience.

Pocahontas rises to her feet and swiftly runs the thirty feet to the center of the Great House. She hurls her small body upon Smith's, wraps her arms tightly around him, and lays her head over him. Everything stops in a great tableau; only the smoke spirals upward through the roof. Then it is known and a great wail goes up among all the people. They are thanking the spirits, and they begin to dance.

Pocahontas is the one who sings out and throws herself over the captive, effectively signaling that he will be transformed while still in the flesh. Although she is a very young Beloved Woman, the presence of the swansdown in her hair means that in the eyes of the council she is of sufficient age and knowledge to be entrusted with this arduous and dangerous task. It seems the manito and the *powa*-currents agreed. Pocahontas becomes something like the god-mother of this new-made Powhatan. Soon enough her commitment will be tested to the full. The ceremony has come to its end. As each person leaves the Great House, he or she gives another gift of tobacco to the fire.

The following day, a meeting with Wahunsenacawh, the great Powhatan, and John Smith is arranged out beyond the main village, in a small hut. There, the *mamanantowick,* Great Man, imposing in his full regalia, a cloak beaded with wampum shells to represent each member of the alliance arranged in concentric circles, tells Smith his new name, Nantaquod. The Englishman heretofore known as John Smith becomes a "son" of Powhatan, *weroance,* or subchieftain. The tribe he is to lead, composed of his fellows at the fort, is known as Capahowasick.

As the Beloved Woman who possessed the *powa,* deciding who would live and who would die, it was Pocahontas, not the priests or warriors or even the Powhatan, the shaman-emperor, *mamanantowick,* who determined the fate of the tribe, reflecting the significant power women held in numerous Native American nations grounded in what ethnographers call "mother right" social and political structure, in which identity and inheritance derive through the maternal line.

Among the Haudenosaunee (misnamed "Iroquois" by the French), for example, each clan was headed by an elder woman. Such women routinely made the decision of life or death. These female clan chiefs, or *gantowisas,* were the female heads of state who directed all of the activities and ceremonial calendars of their respective communities. Among the Tsalagi ("Cherokee" in the English version), the women who hold the power to decide matters of life and death are certain initiated medicine women known as Pretty Women or Beloved Women. Their office is a combination of heredity and medicine ability, both honed by long training and superb discipline among those so burdened.

Among the Powhatans, women had a deciding say in national policy and action. They also owned the great fields of corn, squash, beans, and other staples. They distributed all food and goods, including what was garnered by men in hunting and fishing. They designed and built the dwellings and the lodges for gatherings. It was natural and logical for the Gift of Life and Death to be bestowed on a woman, a Beloved Woman.

The Beloved Woman appears and reappears in a continuum to wield her powerful Gift of Life and Death, as is evident in the story of the Beloved Woman Weetamoo, a Northeastern Algonquin *sunksquaa* who lived after Pocahontas's time. In 1675, Weetamoo's ally the Wampanoag chief Metacomet, known as King Philip by the English, expressed his resentment toward the English settlers for encroaching on his tribal lands and treating his people disdainfully. In 1676, the English waged a horrific war against the Native Nations of the region, known as "King Philip's War," resulting in a rash of raids. In one of these battles, at Lancaster, Massachusetts, the Wampanoag and their Narragansett allies took several settlers captive and held them for ransom; among these captives were Mary White Rowlandson, the wife of a Congregationalist minister, and her three children. It was

Weetamoo, known as "Empress" or "Great Queen" to the English, who exercised her Gift of Life and Death, ruling that Rowlandson should live. Mrs. Rowlandson and her two surviving children served as servants among Weetamoo's people until her rescue some years later before the end of the war. Weetamoo was a formidable war leader who, along with the famous King Philip, directed her people in battle until the very end.

The struggle was protracted, raging up and down what is now much of New England. At one point the assembled Native defenders of their homelands and way of life came close to repelling the invaders for good. Despite some successes, the beleaguered *sunksquaa* and a handful of her fighters fought their last battle. Fleeing toward the Mattapoissit River, where she had spent many happy childhood days with her family, Weetamoo took a small canoe in a doomed attempt to cross to the island there. The pursuing English found her body floating on the stream. She is believed to have drowned. So fearsome was this *sunksquaa,* this queen, that the English took her head and impaled it on one of the high poles of their stockade—an honor that was usually saved for their most hated antagonists.

Smith, in the account of the event at Werowocomoco written seventeen years later, wrote that Pocahontas's gesture indicated her willingness to lay down her life for his. He believed that she was empowered to save him because Powhatan, whom he wrongly identified as her father, doted so greatly on her that he granted her wish. Of course, Smith's interpretation is entirely English. He was unaware of the true nature of the *huskanaw,* ritual death and remaking, of its purpose in the world of the *manitowinini,* or its broader implications in both Dream-Vision and political terms. Being a "new man," Smith was singularly incapable of seeing the greater dimensions of the pattern transforming all around him. While he understood that he was of major importance in this transformation, his understanding of his role was limited to grandiose proclamations of derring-do and tales of intellectual and mechanical superiority.

Smith's explanation was conventional in the English world of his time. Any man finding himself held prostrate by brawny men with fierce expressions while being threatened with a sharp implement that must have looked to him like a nightmare executioner's ax would have been quite certain he was about to be beheaded. Had he been in England, and such was his plight at the command of the king, it would have been an accurate assessment of the situation. In England, any female with such power must be the daughter of a king or emperor, and that ruler's special favorite. And the only reason she would dash out of a press of young women in the midst of a formal and very solemn ceremony and fling herself over the body of the man about to be beheaded would be an act born of passionate love at first sight so intense that her doting father would put down his tomahawk and suffer the captive to live.

But Smith was in the *tsenacommacah* among the *manitowinini,* and what was taking place was definitely not English. The office of Beloved Woman held by Pocahontas, combined with her particular Dream-Vision, which it seems was the factor that propelled her from the assembled girls who were also adorned with white down to throw herself upon the chosen Nantaquod in the proper manner, determined his fate. The decision, as a matter of tradition, would be obeyed by the Grand Council and all concerned. So it was that during the Great Feast, the Nikomis, John Smith was "remade" as an Indian, named Nantaquod, and designated *weroance.* Given all the circumstances surrounding his rebirthing, his selection was more about sacred prophecy and timing than politics, military might, or superior moral achievement. The Feast of Nikomis that winter was, to the Powhatans, the *manitowinini,* more significant than any previous except for the first Nikomis. Through that great ceremony, they let the manito know that they, the Powhatan people, were prepared for the great changes that lay ahead—changes that signaled the creation of a new people. They entered actively into this transformation of their world in harmony with the manito in a *powwaw* where they remade John Smith into Nantaquod, the Englishman of the Royal Virginia Company there in the *tsenacommacah* into the tribe named Capahowasick, which was a large parcel of land encompassing the peninsula where the English were establishing their latest attempt at a North American colony.

OLD MEN'S TALES

As it has been recorded in history, film, fiction, poetry, and biography, the story of Pocahontas is largely a story about the heroic John Smith and the survival of a hardy band of English Christians who came to the *tsenacommacah* ("Virginia," as the newcomers named it). They came to bring civilization to the savage, Christianity to the heathen, and to light the flame of personal liberty, democracy, and the American way of life.

They were aided and abetted in this noble enterprise by a single Indian maiden named Matoaka, but usually called by her nickname, Pocahontas. This "little wanton," as some translated her nickname, sided with the bearded strangers despite the king, her father, Powhatan's anger. She remained loyal to the strangers despite their depredations against her own people, even despite the cruel rejection from Captain John Smith, the man she loved so truly. It is a story told and retold; its outlines are as familiar to Americans as our ideas about the American way of life.

While recent scholars are honest about the depredations against the people of the *tsenacommacah,* the *manitowinini* (Algonquin people of the Atlantic seaboard and Great Lakes), the skeleton of the romance, that Pocahontas loved John Smith and thus threw in with the English against her own people, remains largely intact.

Pocahontas was born around 1597. There is conflicting opinion about the exact date of her birth, which has been lost as so much of the *manitowinini*'s knowledge was lost. The estimates of her birth date, as well as most of the other historical information we have about her, are based largely on John Smith's own testimony. As a kind of scholarly backup to his narratives, indigenous customs and styles are offered, as are some parallel histories by others who were with the original mission into *tsenacommacah* or visited as early reporters.

Girls who were prepubescent dressed in an easily identifiable way, and Pocahontas was dressed child-style when she first visited "James Towne." Smith estimated her age at their first encounter as around eleven or twelve, and other scholars have based the date of her birth within parameters of both his assessment and the testimony of her mode of dress, detailed in the official records of the Virginia Company, which was the name of the English-sponsored venture, into account. The date commonly given as her year of birth makes her as old as possible given her clothing and hairstyle; however, Smith's testimony suggests that she was around eleven years old, or perhaps a slightly built twelve. Since the period of menarche, when child morphs into woman, would have begun by age twelve, the date of 1596 is probably a better estimate than the more common year, 1595.

The child Pocahontas entered the stage of history in January 1608, when Smith, exploring the upper reaches of one of the rivers that ran into the Powhatan (renamed the James River), was taken prisoner by a squad of Chickahominy men, led around from village to village for some weeks, all the time well fed and, had he known it, honored. Their evaluation of him completed, the Powhatan Alliance members held a great ceremony at Werowocomoco, the royal and ritual center of the alliance, where they adopted him as one of its members.

John Smith was remade as an Indian man named Nantaquod. It was the council's wish that he would serve as *weroance* for his tribe at Capahowasick. The place they designated was more or less the peninsula where the English were building James Fort. Near the mouth of the Powhatan River, at the confluence of what are now called the James and the York Rivers, James Fort or "James Towne" was located forty or so miles above the open bay, where they might avoid discovery by Spanish ships. By adopting and initiating Smith, who the Great Council saw as the English's major leader and fighter and wizard—or *mamanantowick*—they thought to defuse an alarming situation.

For his part, however, Smith thought they were about to execute him. He believed that he was saved only because the child Pocahontas flung herself, shrieking, upon his prostrate body. In that position she entreated her fierce father, the "greate king Powhatan," to spare Smith's life.

According to Smith, Powhatan, the great king, unable to deny the daughter whom he held closest to his heart, called off the execution. Smith was not beheaded, and soon enough returned to Jamestown, escorted by several Powhatan warriors. Over the next year or so, Pocahontas often visited the English settlement, disporting herself around the houses with a few English boys who had been pressed into service in England.

During particularly hard times, when the English were going hungry for lack of ability—and inclination—to grow their own food, she appeared with a dozen or so helpers bearing large amounts of corn and other edibles. On another occasion when it was rumored up and down the river that Powhatan was going to massacre Smith and a contingent of men who had traveled upriver to receive a consignment of food from the Powhatans, she came under cover of night and warned Smith that her father planned to slaughter them all the following day. Alerted by the warning, delivered by Pocahontas at great peril to herself, Smith made sure that the food the Indians provided for the English the next day was tasted by the Indian hosts. Smith had heard from other informants on his journey upriver that Powhatan was going to poison them, and as he believed their own weapons sufficient for their defense against a frontal attack, he felt that his precautions saved the expedition. He acknowledged Pocahontas's contribution to the mission's survival as a determining factor in the English party's survival.

Sometime in 1609 Smith's embroilment in factional disputes among the members of the Jamestown company gained him several powerful enemies among the Virginia Company's local leadership. Receiving word that his enemies were about to return to the area from England and that one of his greatest adversaries had been appointed governor of the company at Jamestown, he determined to get some land from one of the local Indians, a friend. He secured the land, but was wounded in a freak accident when some gunpowder ignited on his ship, wounding him so seriously that his worried friends returned him to England, where he could receive the medical care he had to have in order to live.

In the space of less than two years, the Smith-Pocahontas relationship began, strengthened, and was abruptly severed. After his departure, because the conflict between the local people and the interlopers had heated up dramatically, Pocahontas no longer visited the English settlement. It is believed that Powhatan forbade her to visit, but as these were the years when she transitioned from child to woman, a period that is always heavily restricted in indigenous communities, it may be that menarche was the real reason she was not in evidence after Smith disappeared.

However, in 1612, three years or so after his departure, she came aboard an English ship captained by one Samuel Argall (spelled "Argyll" at the time). According to his own account, Argall had worked a deal with a local chieftain by the name of Japazaws and his wife to get Pocahontas aboard the ship under a pretext. For their trouble Argall agreed to give them a copper kettle and some other small considerations. Copper being a metal much in demand among the fashionable Powhatan of the time, his offer was readily accepted.

The plot went as planned, and when Pocahontas was safely aboard ship, being entertained in the captain's cabin, Argall cast off. He would send the older couple ashore downstream, and sail on down to Jamestown. Pocahontas had been abducted by the English and was to be held until her father ransomed her by returning some English weapons he had custody of, delivered a large shipment of corn, and released a couple of English boys who had been given into his custody, as was the custom during more peaceful times.

A message was sent to the "Greate King," and although he agreed to the terms, he failed to make good on his promise. Pocahontas went unransomed. She was taken north to Henrico, a new settlement a few Englishmen had built, complete with a church and pastorage, where a planned boarding school for the purposes of Christianizing and civilizing local *manitowinini* was under construction.

After being dressed as befitted a decent English Christian woman of the early seventeenth century, she was instructed in the beliefs of the Christians and baptized. Her baptism took place in 1613, and in the following year she was to marry another Englishman, a hopeful would-be planter of tobacco named John Rolfe. Rolfe was one of the three Englishmen closest to her during the time of her reeducation.

A small expedition was equipped and made its way to the village where Powhatan was then in residence. He was duly informed of the young woman's impending nuptials and gave consent readily enough. Shortly thereafter Pocahontas, who in giving up her birth name Matoaka had taken the Christian name Rebecca, became Lady Rebecca Rolfe. Her title, lady, came not from her husband, who was a commoner, but from her high status as royal daughter of a man the English had come to regard as "Impire" (emperor) as well as "greate king."

In 1615 Rebecca gave birth to John Rolfe's child, a son whom they named Thomas. By then, the local Virginia Company leadership, feeling an urgent need for an infusion of capital, set sail for England in 1616. They were prepared for some serious fund-raising, having taken aboard Lady Rebecca (as she was known among them at the time) and a party of relatives to serve as her retinue, her son, and several thousand pounds of the tobacco Rolfe had developed, planted, harvested, and cured with the advice, counsel, and effort of his wife's relatives.

The voyage was a resounding success. Rebecca/Pocahontas soon became the toast of London, and when her health began to fail she spent a delightful summer in the country a few miles west of London on the great estate of the Lord of Northumberland. The estate, Syon House, was located on the banks of the Thames. It was favored by the queen and her court as a summer resort, and Rebecca/Pocahontas' recovery there was spent in long walks along the river, picnics in the woods, and evening entertainments both lavish and filled with excitement, as any manner of games, sports, and dramatic dances known as masques were the extravagant and frolicsome queen's delight.

At the year's end the Jamestown contingent made plans to return to Virginia. A date at the royal court at Twelfth Night festivities in early January and extremely high winds prevented their departure until the beginning of March 1617. Lady Rebecca (Pocahontas), who had recovered over the summer, began to suffer decline in the winter.

In late February 1617, as they lifted anchor at London and set sail down the Thames for Gravesend, a small port where they were to rendezvous with a third ship expedition, she was nearing death. Hurried ashore at Gravesend where, it was hoped, she might receive medical attention that would save her life, she died. She was twenty or twenty-one years old.

John Rolfe left his son in the keeping of his brother, a hastily arranged transfer that took place at the party's last port of call, Plymouth, England. He returned to his plantation and continued developing his brand of Virginia tobacco, which soon became a best seller wherever English merchants traded. Although he died four years after his return, his legacy as the father of tobacco—and the plantation system with its rapid shift to a slave-based economy—long survived him.

Meanwhile, John Smith developed into a writer of some skill. It is from his *True Relations,* accounts of his voyages and adventures, along with a few other contemporary accounts, including records of Virginia Company business affairs, that the bare skeleton of the story of Pocahontas is known. It is an English story. It is a story kept by businessmen and adventurers. It becomes an American story, base narrative of a nation where "the state of the nation is business," as President Coolidge so aptly informed us. Despite its origins, it becomes a romance, lending a mystique to those original corporate executives and their struggle for power within the company that would be otherwise lacking.

More, the Anglo-American story of Pocahontas has been told as a woman's story, because women's adventure stories are traditionally cast as love stories in the folk and popular narratives of England, the Continent, and, following these antecedents, the United States.

So Pocahontas entered Western history; the Beloved Woman, shaman-priestess and eventual *weroanskaa* (female leader) of the Powhatan Alliance. She was a woman of many roles and heroic stature, and her four names contain her life's history. Her familiar name, the one she is most known by, was Pocahontas. Her clan or personal name was Matoaka or Matoaks. Her sacred or priestess name was Amonute. Her Christian name was Lady Rebecca Rolfe. Thus began the great ceremony that would lead to the formation of the largest and wealthiest nation the world has yet seen.

In the short span of her life, which was a bit more than twenty years, she would set in motion a chain of events that would ensure the dominance of the *manito-aki* in global life and affairs, usher in a period of terrible decline for her people, liberate the starving and miserable peoples of Europe and beyond, and introduce to a world awakening from feudalist absolutism the idea of egalitarianism, personal responsibility, and autonomy and the initiation of peaceful methods as a way of negotiating national and cultural differences. She would be involved in a great world change in these ways because it was the role of a Beloved Woman to do these things, and, because it was a time of vast change, it was the responsibility of a particularly able Beloved Woman to do so. That woman, as the manito seem to have decreed, was Pocahontas. Among the several roles she filled, the primary one may have been her role as Beloved Woman. Particularly important during times of conflict between the community and its external adversaries, a woman who held this office would find herself called on to decide whether any captives—leaders and warriors alike—would be executed. Her decision was final; no man or woman could override it. The universal symbol of office for such women was one or more white feathers. In the case of the *manitowinini,* this badge of office would have been seen as connecting to the sacred story of Sky Woman, whose plummet through the void was arrested by waterfowl—symbolized by white feathers or, on prepubescent women, by white down.

Because Pocahontas is always depicted with white feathers, the major symbol of the office of Beloved Woman, and because of the role she played in the ceremony during which John Smith's fate—and that of his fellow travelers—was decided, and because he specifically mentioned that the girl who saved him had her hair adorned with white down feathers, we can safely identify her as one who held the office.

In historic times there were fewer Beloved Women than it seems were present during Pocahontas's lifetime. This is because, or so one supposes, this was the time before the precipitous decline in the population of Algonquin and other Native Nations in the Southeast. These were, however, times in which conflict was increasing and strangers were seen, or reported, throughout the regions where the office of Beloved Woman was common practice. In turn, this wide-spread sense of threat was dramatically intensified by prophecies of imminent doom well known among them. So, for a variety of reasons, Pocahontas was not the only Beloved Woman (or Beloved Woman in training) among the Powhatans. Significantly, she was the one who, for whatever reason, flung her small body over Smith's and in that gesture determined that he would live.

THE POWHATAN *TSENACOMMACAH*

No man "understandeth what Virginia is."
John Smith

Occupying a region in the Virginia tidewater region, the thirty-five-plus tribal nations of the *tsenacommacah* lived in villages and towns located on or near several rivers. These mostly drained into the Powhatan River (renamed the James River eventually when the English dominated the region). This great river in turn drained into the huge Chesapeake Bay.

The thirty-five or so tribal nations that subscribed to the Powhatan Alliance were the southeastern branch of the *manitowinini* (Algonquin, Algonkian, Algonkin). Among the members of the Powhatan Alliance were the Powhatan, Pamunkey, Mattaponi, Nansemonds, Appomatocs (Apamatuks), Paspaheghs, Arrohatoc, Youghtanund, and a number of others. John Smith's map of 1612 shows thirty-six "king's houses," or capitol buildings, in the greater *tsenacommacah.*

He includes 161 villages, towns, and hamlets, whose councils were united within the Powhatan Alliance. Most of those tribal states had entered the alliance after the Great Dreamer took office. The English believed that it was the Great Powhatan who, by cunning, cruelty, luck, and much marrying, expanded his sphere of influence from a base group of six tribes whose lands were within a fifty-mile radius of present-day Richmond, Virginia, to a nation that covered most of the tidewater area.

The greater *manitowinini* is a specific linguistic and cultural group, numbering in the hundreds of thousands, at least, at the time of European contact. Their language family was one of the largest and most widespread of North America. Their greater population ranged along the Atlantic seaboard from Labrador to the Great Lakes, along the Mississippi, and to the Carolinas. A few northern plains Native Nations, such as the Blackfoot and the Pawnee, are of *manitowinini* (Algonquin) stock.

The *manitowinini* were the dominant civilization in North America when the European voyages of discovery and great westward migrations across the Atlantic Ocean began in the sixteenth century.

There are various theories among both modern scholars and Native historians concerning their origins. At present the theory of choice among most is that the *manitowinini* originated along the northern Atlantic seaboard, perhaps in the region of Nova Scotia. They may have been the local people whom the Vikings encountered in Newfoundland in the Middle Ages. From that region they spread out, moving west and south. In the main they retained their close association with large bodies of water, only in a couple of instances settling in areas remote from the Great Lakes or the Atlantic.

However, the Western *manitowinini* (Anishinabeg) have another version: they hold to their ancient belief that they originated near the western shores of Great Lakes, particularly Lake Superior and Lake Michigan. This narrative holds that they moved east, north, and south from this area of origin—implying that they did not migrate east to west and north, which was the migration path of the English, and has been the pattern of the United States, but migrated in a pattern of the opposite direction.

A *MANITOWININI* ORIGINS NARRATIVE

One of the versions of the origins of the *manitowinini,* given to the English physicist David Peat (who sometimes collaborated with David Bohm) and recounted in his book *Blackfoot Physics: A Journey into the Native American Universe,* comes from one of the western *manitowinini* peoples, the Ojibwaj (Ojibwa, Ojibway).

"Native America has its own stories of origin, that of the Ojibwaj, for example, going back to the last ice age." Peat tells us that his being told even this small fraction of the entire origin cycle was controversial among the local people. The elders argued over the propriety of sharing their narrative because the teachings instruct us that "the stories are sacred and must never be passed on to outsiders." Evidently, the eventual consensus was to go ahead with the parts that he recounts in his book because those who argued that "the time has come to speak openly and share their knowledge" prevailed.

Peat tells us that "the sacred stories of the Ojibwaj peoples were recorded, using a symbolic language, on birchbark scrolls. As these scrolls age and begin to disintegrate they are meticulously copied. . . . The history of these stories can also be found carved and painted on rocks . . . ," an assertion that flies in the face of the common assumption that Native people lacked writing. He continues with his account:

> The knowledge of the Ojibwaj peoples is preserved and passed on within the initiation ceremonies of the Midé-wewin, or Grand Medicine Society [in which Pocahontas was trained], and, at each level of initiation, the teachers take whole nights to complete their work. Some of the stories speak of a giant of ice, the [Windigo], of the growling noises he [makes], and of how The People [Anishnabi, Anishinabeg] would face death if they approached too close.
>
> One story suggests that the Anishnabi lived by the shores of the saltwater for so long that they began to forget their origins. Then one day a *megis* shell appeared above the water to remind The People of their origins. The Anishnabi followed

the *megis* shell on a journey east that took them along what we now call the St. Lawrence River, into the Great Lakes, and on to the north shore of Lake Superior.

The birchbark scrolls are quite specific about this phase in the Anishnabi ancient migration. Symbols on the scrolls can be identified with landmarks such as waterfalls and islands in the Great Lakes region. It was from this area that the Ojibwaj spread out, taking with them their history and teachings and inscribing and painting it on rocks and scrolls.[4]

While this version belongs specifically to the Ojibwaj Nation, Canadian branch, it is substantially the same one that belongs to the Lenape, a *manitowinini* nation very closely related linguistically and historically to the Powhatans, Pocahontas's nation.

LIFE IN THE *TSENACOMMACAH*, CA. 1600

As for Tidewater Virginia—that coastal strip extending back some one hundred miles to the western fall-line, to the Appalachian piedmont—[it is] a land in thrall to the sea, subject to the ebb and flow of the tide through the estuaries, deep into the interior; five of the rivers—the James, the York, the Piankatank, the Rappahannock and the Potomac—spread out like the five fingers of a giant hand, thrusting into the land and dividing it into verdant strips and peninsulas.[5]

Like any national community, the people of the *tsenacommacah*, which included most of the Algonquin population in the region and extended over most of the tidewater area, had their own cultural style. It included their dialect of greater Algonquin, their mode of dress, eating implements, recipes, healing protocols, architecture, household goods, adornments, and religious practices and beliefs.

Many scholars believe that the name of the entire allied group of individual tribes, Powhatan, which was also the name of one of the member-states of the alliance, was taken from the name of some large falls at the northern boundary of the *tsenacommacah*. That place-name meant "the central source of power dreaming." It seems more likely that the falls were one of those power places whose energies, for whatever reason, brought about altered states of consciousness; hence its name. The word *powa,* which forms the base of the name of the falls, Powhatan, generally signifies otherworldly or paranormal perception and ability. As Peat reminds us, "Within indigenous science the voices and images of dreams are not symbols of the unconscious but aspects of a reality that is far wider than anything we assume in the West."[6] As his comment indicates, it is difficult to describe in modern terms the state of consciousness that is more or less common, if not ordinary, among the People of the Dream, the Powhatan. Suffice it to say that such states were integrated into awareness and expectation. Living in a Dream-Vision world was not abnormal; on the contrary, it was a desired state, probably reserved for certain individuals belonging to higher-status clans. However, there was a ceremony for boys called the *huskanaw* that allowed the identification of those with inborn gifts from all classes. Whether a similar institution existed for girls is unknown, though one can assume that those who moved through the various levels of the *midéwewin* would show such gifts early on.

Words that were based on the root *powa* include *Powhatan,* which can be translated People of the Dream-Vision. The word *powa* was also used as honorific for the most powerful dreamer (as a man's title *powhatan* means "he who dreams"). During the period of Pocahontas's career among the English, the Chief Dreamer was a highly gifted fighter, strategist, politician, and ladies' man named Wahunsenacawh. The usual title for men with such a wide range of excellence was *mamanantowick,* but so gifted was Wahunsenacawh, his abilities outstripped those of such great men, that he was given the honorific title or name *Powhatan.* The root, *paw, pow,* or *po,* occurs in words with similar connotations, words such as *apowa,* "Dream-Vision"; *pawwaw, powwow,* "we Dream-Visioners Dream-Vision together"; and *nepowa,* "I Dream-Vision." *Paw* (or *pow,* sometimes shortened to *po*) occurs in words that possess some degree of *powa* or that refer to it. Thus, the Potomac River, which runs through the capital city of the United

States; and *pocahaac* (bodkin, awl, and other long, narrow objects), a word that is related to the word *Pocahontas*.[7]

Pow-a, paw-a, or whatever variant spelling is used, is a complex word meaning the connection with paranormal or supernormal perception and ability. It was, the *tsenacommacah* believed, manifest everywhere. This necessarily included daily life and "material culture"—that is, the "things" humans in a given social system possess. *Powa* accrued to clans, some having more, some less; it also accrued to tribes. The Mattaponi, Pocahontas's tribe, was a powerful tribe within the Powhatan Alliance. This power was a consequence of the degree or "amount" of *powa* inherent in "Mattapon-ism," so to speak, not because it was generated by humans of earlier times or in the present, but because it was inherent in the "gestalt," the field of being, that was identified as Mattaponi. Thus, those who were born within that tribe shared in a degree of Mattaponi-*powa*.[8] The social and material consequences of *powa* accumulation were shown by a number of social indicators.

Wealth was a matter of clan affiliation as well as tribal membership. A class system operated, if English reports are to be credited, and the differences were evident in apparel as well as in workload, spiritual and ceremonial status, and the number of individuals a given man or woman—or child—commanded. It was also integrally connected to one's village and tribe. Those that were smaller were less wealthy because trade was very active among them, and those communities with more to trade had greater personal, village, and clan wealth to distribute.

The *tsenacommacah,* as the Powhatan group referred to their lands, covered a large land base. Stretching from the Chesapeake Bay northward to the Rappahannock River and south to the York, they were bounded on the east by Sioux and Cherokee settlements. Theirs was a heavily wooded territory, rich in hardwood growth, which meant good hunting; rich soil; and usually good climate. It yielded such sufficiency of food and other items such as tobacco that the people generally had plenty to see them through the winter, village by village, with enough left over to sell large amounts of both to the English.

Because of the bay and the extensive rivers, estuaries, creeks, and swamplands, a plentiful supply of fish, clams, mussels, and other water-compatible edibles was assured. There was a plentiful supply of reeds for weaving mats, bark for making containers and siding, plenty of trunks and branches for the support structures, and roots for root clubs. There were generous supplies of uncultivated foods such as berries, green vegetables, fungus and root plants, along with herbs for healing, flavoring, body paints and dyes, purification, and consciousness altering for initiations and other sacred events. This bountiful land offered up more than sufficient quantities of feathers, shell, and skins for clothing, adornment, footwear, and trade. Although cultivating, hunting, fishing, and gathering required labor, theirs was usually a world that richly supplied a lifestyle more given to holidays (holy days) and leisure than endless scrabbling for bare survival. Because the land was so bountiful and they so knowledgeable about cultivation, preservation, hunting and tanning, jewelry making, medicine, and the like, they engaged in a thriving trade between villages and communities outside the *tsenacommacah,* the Powhatan Alliance.

In the winter months the skins from which their clothes were fashioned retained the fur or hair of the animal it came from, while summer and ceremonial garb was likely to be of soft, tanned hide, sometimes white buckskin. In his account of one meeting with Pocahontas, John Smith describes her outfit as a cloak, moccasins, and leggings make of soft, white deerskin, with a dress to match.

While women of the alliance usually wore a garment constructed like a copious apron, they also wore a similar garment that was clasped over one shoulder, covering one breast. Men wore similar garments, although they favored a plain loincloth and sturdy low moccasins. Drawings made by English travelers such as John White, governor of the lost colony of Roanoke, indicate that there was a wide range of style and material for clothing, and apparel depended on one's status, one's field of specialization, or the occasion. Work or everyday dress was distinct from dress for more formal occasions.

By all accounts, the coastal Algonquins of Pocahontas's time were a handsome people, "comely," well formed, generally taller than the English. Their faces were broad, and men and women decorated their visage with tattoo marks. Most were graced with heads of thick hair—though the men had sparse facial and body hair; Smith describes Wahunsenacawh's beard and mustache as "exceeding scarce" and wispy. Seventeenth-century Algonquin men were consistently drawn without facial or chest hair. Tweezing unsightly hair on face, scalp, or chest was common enough, although men—or women—did not have a lot of body hair. Removal was usually accomplished with a pair of clamshells pinched together to pull the offending hair.

Men were likely to wear their scalp cleanly plucked, with a top-knot and long lock hanging from the left side, indicating their status as warriors. Shamans and priests, like *weroances,* had identifying hairstyles, as did boys, girls, and women. The children's hair was kept shaved on the sides and worn otherwise long, much like the men's. Women wore their hair with bangs and most often bundled into a chignon, although they also left it long and swinging or simply clasped at the back. Adult women did not usually shave or pluck their scalps; a full head of hair seems to have been a mark of full-adult status.

Adults of both sexes were identified by intricate tattoo patterns not only on their faces but on their wrists and legs. The leg tattoos circumscribed the leg at about the calf. John White as well as painters who followed him left some examples. From these it can be seen that some designs are geometrical—resembling sharp-angled chevrons with small, dark lines and/or dots about a half inch in diameter in a carefully designed pattern. Other tattoos, mostly worn by women, are examples of what ethnographers have designated "woodland design." This pattern resembles the twining and curving of vines, a kind of infrastructural floral pattern. Floral patterns are an enduring symbol used by Natives of the American eastern forestlands. They adorn today's powwow dresses and dance shawls; ladies' dance bags, wide belts, and fans; men's parfleches and quivers; and most of the beaded moccasins, worn by both men and women, that come from that region.

The settlements varied in number of houses and public buildings, ranging from hamlets to towns. The English late in the seventeenth century estimated the overall population of the *tsenacommacah* to be around eight or nine thousand, and the population of all of what is now the state of Virginia as around eighteen thousand. Their numbers provided the *tsenacommacah* with numerical superiority in the region; their nearest neighbors, the Chickahominy, who were not full partners in the alliance, also numbered a few thousand.

Villages consisted of houses and major buildings arranged more or less randomly around a center, where ceremonial and other social occasions occurred. Fields of corn, pumpkin, tobacco, squash, and beans were close to or virtually within the village, giving easy access for tending and harvesting. Women owned the fields and were responsible for the produce, and while the women did much of the gardening, men provided muscle for plowing, fencing, and sitting in a specially erected shaded platform to scare off crows and other shoplifters.

The houses tended to a rounded overall shape. The sides were made of birch branches that were bent to form a hoop over the top of the living space. Smaller branches were interspersed between the larger ones, and sizable posts framed the door. The outer walls were covered with large sheets of birch bark or woven mats—the bark being preferred, and lashed to the branches that were firmly entrenched in the ground. The door was covered with a flap made of skin or tightly woven mat; the interior, walls and floor, were covered with similar mats.

There were large dwellings as well as public buildings such as *quioccosan,* Great Houses (or "temples," as John Smith termed them), similar to the one where Smith was initiated into the alliance; *midéwewin,* Great Medicine Dance houses; and edifices known to the English as "Houses of the Kings," where the skeletons of great leaders were kept. These buildings were of impressive size, being as long as two hundred feet and fifty or sixty feet wide. John White's rendering of one House of Kings shows it as tall as the height of two men, making it nearly a two-story building.

The longhouses, as they are known today among many communities of Algonquin and Haudenosaunee stock, were multifamily dwellings or council halls. Shaped like modern Quonset huts, rectangles with curving roofs, these buildings varied in length and width but were constructed using techniques and materials similar to those used for the smaller houses. In addition to the mats gracing the walls and warming the floor, benches were erected in two rows along the walls. The higher ones, closest to the wall, were usually used for storage, while the lower row doubled as seating and sleeping space. There were at least two hearths in these larger dwellings, circular in shape and located beneath a smoke hole in the roof. There cooking, gossiping, smoking, and group life went on, no doubt noisy and tumultuous. Little ones, older children, and adults of varied ages, genders, and temperaments lodged together. Things might have gotten a bit tense during long winter snows or blizzards, but much of the time the men were out hunting, fishing, fashioning clay pipes and smoking, curing tobacco, and making their smoking mixture, called *pissimore*. They could occupy time just hanging out (a favorite way to pass time, as it still is among modern Indian men). There was time spent helping in the fields—although most of the work of cultivation was done by the women who owned the fields—or sitting watch for crows (they built small huts that were raised a few feet above the fields for this purpose). Men could also spend time fishing, clamming, or going about priestly, governmental, and social tasks, attending village or intercommunity councils, and trading for a variety of goods including sheets of copper.

The women, on the other hand, were responsible for maintaining the day-to-day life-support system of the whole community. Their time was devoted to weaving mats, fashioning various sizes of storage containers of bark and woven sweetgrass, decorating their homes and the public buildings, gossiping, nagging the men, Women's Council meetings, spiritual study and practice, gathering and storing herbs, teaching the complex details of their countless disciplines to younger women and girls, watching the little children, nursing babies, making jewelry—which included polishing, drilling, and stringing pearls and shells—making clothes, tanning hides, transporting meat, fish, clams, oysters, and wild and cultivated vegetables to distribution centers, trading up and down the rivers and creeks for corn, tobacco, and other goods, having babies, taking care of sick people, entertaining elders, preparing meals and feasts, preparing the halls for ceremonies and festivals, and getting some well-earned break time once a moon when they got to retire to the women's house and be waited on by women who were not menstruating. There the time was passed in talking, meditating, and reflecting, planning one's next adventure, project, or costume for whatever big gathering was coming up, and most of all being safely secluded from the tumult of daily life around the *tsenacommacah.*

The English reported that "most of the work was done by the women"—whom they characterized as "drudges," but they also made it clear that drudgery had a great deal to do with status. Women and girls such as Pocahontas, privileged by birth, clan, and destiny, did not engage in any real drudgery. The same could be said for many of the men who would have come to the Englishmen's attention. As the fields, houses, household goods, children, information about food— getting, preserving, preparing, distributing, trading for, and the like— were all in the hands and under the direction of the women, it is hardly surprising that the women also oversaw and took responsibility for their holdings.

Property, household or field, was inherited matrilineally, and families resided with the wife's clan—that is, with her mother's people. On the whole governance was along female lines, with female heads of clans being the primary policy makers. Men were generally the public speakers and implementers of the policies that were socially significant. In spiritual matters the tradition—along with the mood of the manito and the *manito aki,* the spirit entities and the spirit world— decided them. Men and women alike who were well versed in the traditions of their people and who possessed certain faculties, belonged to certain clans, and were given certain kinds of Dream-Visions occupied the office of priest/priestess or medicine woman/man. *Medicine,* a term that seems to have been in vogue among those who wrote about American Indian spiritual matters, is based on a corruption of an Algonquin sacred tradition that is complete with ritual and liturgy. Known as the *midéwewin,* the Medicine Way or Medicine Dance, it is not necessarily concerned with healing the sick. It might be concerned with interactions between the human sphere and the *manito aki;* it might have to do with finding lost objects, lost loved ones, or lost causes. It might be concerned with world-renewal ceremonies, or with keeping the community in harmony with cosmic currents as they swirl around the people, whether human, animal, or plant.

In terms of leadership styles among the *manitowinini,* to say someone "commanded" a group, a tribe, or the alliance—which is the English way to signify a primary leadership function—distorts acceptable leadership modes among the Native people. Terms like "major networker," "central tradition-focus," "connects disparate groups with one another," or "hub" might better identify how Powhatan leaders worked. They tended to show their inclination early in life. Most clans and tribes could identify a leader before a child was six or seven. Pocahontas was recognized as a leader from childhood. The English reported that she was always followed by others, "her wilde band," and, later, other women, "her sisters." She seldom bore burdens, but led those not blessed with leadership ability. To English eyes that reinforced their belief that she was an upper-class child, daughter of someone at least as important as "the greate king, Powhatan." It is more likely that her innate ability singled her out rather than her mother's position (and in a mother-right system the father's identity could have had little to do with matters of primary status). Even after she became the Christian wife Rebecca Rolfe, she was attended by a variety of Powhatans, thought by the English to be her relatives (how family relationships were determined in the *tsenacommacah* differed from the English usage). The only period in which she was unaccompanied by Mattaponi or companions who lived in Werowocomoco was her premarital sojourn at the small settlement outside Jamestown, Henrico.

Even then she was always within call of her clan relatives—which are relatives but not in the sense of a nuclear or even extended family. After her marriage to Rolfe she was given a large parcel of land, a gift to which her birth status entitled her. The English, of course, believed it to be a gift from her father, Powhatan, but that is not particularly likely—or relevant to the ways of the *manitowinini*.

Scholars are aware that Pocahontas was royalty—that is, was in line to inherit alliance leadership, rule by inheritance being the mode in use among the Powhatans at the time. However, there seems to be little scholarly recognition that the position of Powhatan—that is, Principal Chief of the Alliance—depended at least as much on spiritual, shamanic, or magus ability as on clan. This means that while Pocahontas's chances of gaining the office of Powhatan seemed distant if figured in accordance with Western-world kinship systems, it was probably a great deal closer, had she returned to the *tsenacommacah* or Jamestown colony after her sojourn in England. She was a consummate Holy Woman, far more thoroughly educated, trained, and disciplined than a "shaman." Her level of ability was more like that of High Priestess than Native habitué of the *manito aki,* the world of the manito. The very acquiescence of the *matchacómoco,* the Grand Council, to the part she played for over eleven years as liaison between the English and the *manitowinini* testifies to the status they knew she occupied and potentially would occupy when the time came—should their way of life survive until she was of the proper age.

MOTHER RIGHT AND POCAHONTAS

In the *tsenacommacah,* society was organized as a mother-right civilization. In such societies, descent is determined matrilineally. In mother-right systems, the identity of one's mother determines one's family, or, to be exact, one's clan. Clan membership is the basis of personal identity in a large number of Native communities. Sometimes the clan is determined patrilineally, in others matrilineally. The people of the *tsenacommacah* used the latter system.

Among them also, when a couple married they took up residence in the village, or in most cases, in the great house, of the wife's clan.

This living arrangement saw several families living under a shared roof in a building that might be as much as a hundred feet in length and perhaps fifty feet in width. In the big house, both husband and wife were subject to the rule and decisions of the clan head, usually an older woman who was not only older but hereditary head of that clan. The *weroanskaa* (*weroansqua*, *weroanska*, *weroansquaw*), whose word was law, made decisions that were strongly grounded in the ceremonial tradition, whether oral or written.

The children of the union of a particular woman and her husband were seen as the children of the clan, or "longhouse," from which their mother descended and to which she belonged. Male authority was in the hands of the clan matron's male relatives: her brothers, uncles, and sons. They played a major role in village and national governance as well as in family and clan matters. Some researchers have suggested that by the time the English came to Jamestown the Algonquin of the eastern Atlantic seaboard were moving from a matron-centric system to a more male-centered one. Given the reports of the English, who were highly influenced by their own social conditioning and expectations, male authority seemed rampant. If this was the case, the likelihood is that a system one might classify as "avuncular" rather than "patriarchal" was forming, because the tradition of woman-owned lands and houses, woman-determined lineage, and woman-centered leadership was very much in place in the early seventeenth century.

Because the yield of the fields along with that of gathering, hunting, or fishing expeditions was the responsibility of the *weroanskaa,* they made decisions about the distribution of goods and other results of hunting and gathering expeditions, and they made provision for storage of dried corn, beans, and other edibles, and of herbs for healing and other ritual matters. Men owned their own clothes as well as hunting, fishing, and ritual paraphernalia. In the event of divorce the man returned to his mother's clan village, where he remained until he remarried—if he did so.

Men held office in the village tribal and intertribal council by virtue of appointment to office by the leader women. They could be removed from office by a decision taken by a council of *weroanskaa,* including the *weroanskaa* of his maternal clan. Appointment was indicated by the presentation of antelope horns to the honoree, bestowed on him by the clan's leading woman or, from an English point of view, the queen.

WAHUNSENACAWH, THE GREAT POWHATAN

Because of the law of mother-right that was practiced in the *tsenacommacah,* it is impossible to say with any confidence that Pocahontas was Wahunsenacawh's daughter, eldest or otherwise. For the Indians' benefit, or so they thought, the English habitually used the term *father* to signify the leading man. It is no surprise that the Powhatans did likewise, doubtless thinking to keep communications clear. But in recounting the events, at least when referring to the Powhatan chief-dreamer's relationship to Pocahontas, the English became perversely literal, ascribing paternity to him. Because of their misunderstanding, we all think Pocahontas was Wahunsenacawh's daughter—an idea evidently fostered by locals who chatted with one or two English who recorded that the rulers among them indulged in a kind of serial monogamy, keeping the issue whether female or male, and sending the child-bearer off to her village or town of origin. It is not possible at this date to ascertain the validity of this tale, or even its source, English or indigenous.

In the end, given the intricacies of communication between peoples who did not share either a language or assumptions about the nature of reality or propriety, it can be argued that much of the information we have from seventeenth-century English sources is misinformation or, at least, misinformed information.

The English thought "Powhatan" was Wahunsenacawh's name, but it was his title, or the name of the office he held. *Powhatan* meant "principal dreamer," meaning that Wahunsenacawh was the most powerful dreamer among the people of the dream, the Powhatans. Most dreams of the kind courted in Powhatan society resulted in real effects in the material world, but Wahunsenacawh's were the most potent *makers* and *changers* of mundane reality, which is why he was the Dreamer of Dreamers, Powhatan.

CREATION STORY OF THE *MANITOWININI*

Every Native autobiography or biography begins with the creation story, includes the "origin" story, and contains all stories or sacred narratives that have bearing on the significance of the life being told. There are several reasons for this practice, chief among them the belief that an individual is known only in terms of the community. The community is seen as not only a few human beings such as one's family, schoolmates, spouse, children, and perhaps colleagues or fellow workers. Instead the community is defined by the supernatural beings and forces that birth it and maintain it, and with whom interaction is a daily occurrence. It includes all the kingdoms categorized as animal, vegetable, and mineral. Community includes climatological, meteorological, astronomical, and geographical features. Community, in Native understanding, is holistic and inclusive; it is not contained within either a specific geographic locale or within a specific temporal one, as both are understood by most modern peoples. It is not confined to three dimensions plus time, as we humans—at least we moderns—appear to be. Affiliation or kinship is one of its major features. This being the case, we can discover the meaning of Pocahontas's life only by knowing the mythic dimensions into which she was born and in terms of which she understood the meaning of all life and hers in particular.

The *manitowinini*'s central creation myth, one of the most powerful and therefore most sacred of their narratives, is the myth recounted earlier, the Woman Who Fell from the Sky. The place she came to rest, the back of a Great Turtle, some say became the earth we inhabit, known to the *manitowinini* as Turtle Island. However, the actual Turtle Island the *apowa* referred to originally may well be an actual place, geographically quite a bit smaller than the entire earth. There are many such places, many of which are scattered around the traditional lands of the *manitowinini*. They are powerhouses of some kind of force field, said to be a magnetic field by those who study them.

Such a place, known as Turtle Island, is located in a remote part of upper Wisconsin on or near the shores of Lake Superior. Those who live nearby say that it is an ancient earthwork. Perhaps it was constructed to resemble more strongly the *powa* of the Turtle Manito for which it is named, with the builders using geographical features already there as the base, which they modified. There are a number of such constructions around the world, some well known, others just coming to be identified. All of them have some connection with ancient astronomy and what modern people think of as religion.

The world Pocahontas lived in began with the descent of Full Bloom, also known as Sky Woman, to Turtle's back. This event, rendered variously as a mythic narrative and a sacred ceremony, is most accurately understood as an astronomical event. It was a primary case of "standing in a sacred place"—which is the same as "dancing in a sacred time." It was one of those world-changing times, caused by an event recorded in their astronomy as well as ritual, that occurs from time to time, these times perhaps separated by thousands of years.

THE ASTRO-METAPHYSICS OF
NATIVE AMERICAN NATIONS

In the past few decades, scholarly interest in the relationship between astronomical knowledge and social structures among Native peoples has received increasing attention. A field known as "archeoastronomy" is flourishing, and alongside monographs and books, one can locate computer programs that enable one to view the constellations over a period of several centuries in their relationship to particular locations on earth.

In his highly informative study *Lakota Star Knowledge: Studies in Lakota Stellar Theology,* Ronald Goodman tells us:

> Recently we learned of some artifacts which clearly define this mirroring. Research into Lakota stellar theology has added new dimensions to our understanding of how the People generated the mentality for experiencing the sacred. It shows that they felt a vivid relationship between the macrocosm, the star world, and their microcosmic world on the plains. There was a constant mirroring of what is above by what is below. Indeed, the very shape of the earth was perceived

as resembling the constellations. For example, the red clay valley which encircles the Black Hills looks like (and through Oral Tradition is correlated with) a Lakota constellation which consists of a large circle of stars.

They are a pair of tanned hides; one hide is an earth map with buttes, rivers and ridges, etc., marked on it. The other hide is a star map. "These two maps are the same," we are told, "because what's on the earth is in the stars, and what's in the stars is on earth."

The Lakota had a time-factored lifeway. The star knowledge helps us to understand this temporal spatial dimension more fully.[9]

This geological feature of the Paha Sapa, the Black Hills, is uncannily similar in its astrological significance to a similar geological formation in southwestern England known as the Glastonbury Zodiac, although the latter is, of course, a geological mirror of the Western Zodiac. The Sioux were originally located in what is now the American Southeast, perhaps in the region between Mississippi and Virginia. They migrated north along the Mississippi River to the eastern side of the headwaters of the Missouri, where it and the Mississippi converge, around 800, European time. In astronomical terms, relevant to locations of significant constellations and significant stars, their original location was in the same quadrant in which the homelands of the Powhatan Alliance were located several centuries later.[10]

Although the two groups were from separate language stocks—Siouan and Algonquin—the location had a great deal to do with how they constituted their sacred reality. The stellar configurations in the south differ somewhat from those of the northern Midwest; this means that the religions would differ to some extent; but while that is always true, there is a certain agreement about cosmic realities among the peoples of the Western Hemisphere. Indeed, it would seem that there is a certain agreement about cosmic realities among the peoples of the earth. Archeoastronomers have devoted a great deal of study to another star-chart cosmological system, that of the Skidi Pawnee.

There are a number of points made by various scholars concerning the Skidi people's star-chart hide. While much of it is obscure to contemporary astronomers because the Native system of naming constellations and stars is not based on Greco-Roman or modern astronomical designations, some features are discernible. For our purposes in understanding the kind of mentality Pocahontas possessed, one particular feature is worthy of note. Among the stellar groups they identified is one, known to them as "The Swimming Ducks," that bears significantly on the Sky Woman narrative.

This stellar configuration has been identified on the Western star chart by the astronomer Forest Ray Moulton as the stars Lambda and Upsilon Scorpio. The two stars form the stinger in the Western constellation Scorpio.

So, I would argue, the fall of Sky Woman was an astronomical event. It was also a metaphysical event, an event that occurred in the *manito aki*. So a powerful new place came into being, one that worked its way into the three-dimensional, time-ordered solar system we recognize. The way the astronomical event, whether astral, or cometary, was recognized was by its sudden appearance among the Swimming Ducks. While the Pawnee are presently situated pretty far to the west of the main body of the Algonquin group, their chart might be a version or the template of what the Algonquin-cum-Haudenosaunee ("Iroquois") creation myth alludes to. One ethnolinguist has noted that that among the northern Algonquin, the word often translated as "ducks" is a generic word that could mean any waterfowl, such as swans, geese, or loons. Because a Beloved Woman is always marked by the presence, in her hair or hands, of swan feathers, the connection between Beloved Women and the cosmic event the narrative of Sky Woman refers to is clarified. A Cherokee scholar has informed me that in her language the word for swan can also mean goose.[11] While the Cherokee were in the sixteenth century near neighbors of the Powhatan but not of Algonquin language stock, it is reasonable to assume that Sky Woman, whatever stellar phenomenon the narrative refers to, was "caught" in the stellar system designated Swimming Ducks—waterbirds—by the Skidi Pawnee. It is further significant

that, while the Pawnee are of Caddoan language stock, their ceremonial system is closely related to that of the Blackfoot Alliance. For at
least two hundred years, the Blackfoot have lived in what is now the
province of Alberta, Canada, and in Montana. Their astrological systems are connected to the Sky Woman narrative, as is recorded at a
place in southern Alberta known as Writing On Stone. This ancient
sacred site, located in a place currently known as Police Coulee, sits
along the Milk River. The Milk flows into the Missouri, which flows
into the Mississippi. Somewhere along the line the Milk converges
with the Ohio, in the area where are located a variety of mounds,
earthworks that appear to have native zodiacal connections, as well as
hobbomak.

Studies that explore the astrophysical foundation of Native ceremonial life and the social structures that mirror it are providing a very
different picture of the world Pocahontas knew. The Pawnee hold
that they came from the stars. Whether the Powhatan held a similar
belief must remain unknown for a time, but given the migration narrative of the northern Algonquin, and knowing that the Powhatan are
offshoots of their cousins to the north, it is likely that they held the
same view of their origins. In any event, as the *manitowinini* trace their
origins to the descent of Sky Woman to Turtle Island, they must also
see themselves as descendants of star nation people who lived somewhere the Oral Tradition now names Sky World.

One of the tenets of Algonquin religion as practiced by many of
that stock is that when the event signified or heralded by the coming
of Sky Woman to Turtle Island recurs, as it does on a regular basis, the
world is on the brink of massive change, or renewal.

Given the situation along the Atlantic seaboard in the years
stretching from at least 1492 into Pocahontas's time, it seems clear that
the Powhatans, like their Mayan–Aztec neighbors to the south across
the Gulf of Mexico, knew that the time for world change was upon
them.

Pocahontas was the blossom of a cycle of world renewal that was
kicked off for her with her *apowa,* and sealed when she flung her child
body on the supine form of the strange little man she knew as

Nantaquod. She did her best to help him learn the ways of the *manito aki,* but in the end she failed. Smith left the area for England, disappearing without telling anyone in the *tsenacommacah* where he was going. Pocahontas believed, as perhaps did everyone in the *tsenacommacah,* that Smith/Nantaquod had died. Evidently, while his sudden disappearance entailed a modification in the plans of the *manitowinini* and their manito guides and gods, Pocahontas remained entirely among her people for three years, then went to Jamestown, where she was converted to Christianity and married an Englishman. She discovered that Smith was alive only several years after his sudden disappearance, when she went to England. The adjustments that the council and/or manito made may not have given the desired or optimal results, but by the time of Pocahontas's death she was and would remain one of the best-known Indians the English ever knew. There are still pubs in England, surviving from her era to this, named "The Indian Maiden" or "The Indian Princess."

In order to understand the far-reaching consequences of such a short if glamorous life, knowing who Pocahontas was is primary. For it is clear that she was by birth, vision, training, and circumstance the agent of change—necessarily female because the Oral Tradition leans toward female *powa* as initiating the new movement or "world renewal" everywhere in the Americas. Looking back over the past four centuries since her lifetime, we can see how profoundly the world has changed; not only in its human dimensions, but in every aspect: the earth, the entire biosphere are greatly transformed. The state of human knowledge—of one another, of the vastness of their earth and what lies beyond it, the awesome powers unleashed by science and technology—in all these dimensions and more, this is not the world where Pocahontas lived and died.

A CHEIFF LADYE OF POMEIOOC
(printed 1590)
Engraving by Theodore De Bry
(courtesy of the John Carter Brown Library at Brown University)

Pocahontas / Mischief

Were there two sides to Pocahontas? Did she have a fourth
dimension?

—Ernest Hemingway

She has been called "the first lady of America," "a daughter
of Eve," "a child of the forest," "a madonna figure," "the
non-parallel of Virginia," "the mother of us all," and "The
Great Earth Mother of the Americas."

—Charles R. Larson

Pocahontas was abducted, or so the story goes. How this event came about makes interesting reading, although the English version as adapted by various biographers necessarily leaves out many details that are of significance within the Oral Tradition and that further illuminate Pocahontas's sacred role in the early development of colonial America.

The abduction of Pocahontas, her Christianization, and her marriage have been summarized in *The Columbia Encyclopedia:*

> In 1613, Pocahontas was captured by Capt. Samuel Argall, taken to Jamestown, and held as a hostage for English prisoners then in the hands of her father [*sic*]. At Jamestown she was converted to Christianity and baptized as Rebecca. John Rolfe, a gentleman settler, gained the permission of Powhatan and the governor, Sir Thomas Dale, and married her in April, 1614. The union brought peace with the Native Americans for eight years. With her husband and several other Native Americans, Pocahontas went to England in 1616.

After her encounter with John Smith, at which she sponsored him and, through him, the entire Virginia Company of Jamestown into citizenship within the *tsenacommacah,* she spent the ensuing few years growing up. Surely she continued her studies within the Great Medicine Lodge. In the months following the Great Ceremony, she was a frequent visitor at the English village, where she gained the men's notice by her abandoned play. The practically naked youngster's ebullient cartwheeling about the fort must have left the men bemused. It was such behavior that led many commentators to see her as "wanton," a term some said her nickname, Pocahontas, signified. It was also in this youthful naked exuberance that she first caught the eye of John Rolfe, a widower she would eventually marry.

The image left us by William Strachey, that ubiquitous chronicler of the infant settlement, is vivid:

> Pochohuntas [*sic*] a well featured but wanton young girle
> . . . sometymes resorting to our Forte, of the age then of 11,

or 12, yeares [did] gett the boyes forth with her into the market place and make them wheele, falling on their hands, turning their heeles upward, whome she would follow, and
wheele so her selfe naked as she was all the forte over.[1]

So vivid was his description of the girl-child disporting her naked
"selfe," "privities" covered only by the bit of moss female children
wore—a kind of thong—that it was not published until 1849, though
it was written in 1612.

The well-dressed Powhatan woman of the time—anyone at and
after puberty—wore a much more capacious garment that resembled
an apron. It was made of a length of cloth that draped downward
from the hips, falling over the crotch to a length of maybe ten inches
below the navel. The outfit was sans bodice, a fact that undoubtedly
contributed to the Englishmen's view of the women of the *tsenacommacah,* the larger Powhatan community, as uncivilized (though one
might point out that the customary dress of the English court left a lot
of breast uncovered, as renderings of recently dead Queen Elizabeth
show). Strachey estimated her age when she enjoyed larking about
with the fort's children—presumably all boys—because "[t]heir younger
women goe not shadowed [without skirts] . . . untill they be nigh
eleaven or twelve returnes of the leafe old. . . . But being past once 12
yeres they put on a kynd of semicinctum leathern apron (as doe our
artificers or handicrafts men) before their bellies and are very shame-
fac'd to be seene bare."

Women clothed their upper bodies with generous strands of beads
made of delicately wrought wampum shells, copper, and pearls. The
long strings of beads cascaded like leis, revealing tattoos as they swung
with a woman's movements. Like the tattoos, the number and com-
position of the strands also revealed the woman's wealth and status.
Fashionable women decorated their ankles, wrists, and chins with el-
egant tattoos that announced each woman's clan affiliation and other
clues about her identity, much as makeup, hairstyle, clothing type,
label, and hue announce ours. A young woman's long hair was
likely to cascade over the breasts, presenting a reasonably modest
appearance, while matrons wore theirs more respectably twisted into

a thick smooth bundle clasped at the neck. Many a fashionable young woman on the contemporary scene echoes Algonquin women's dress of the time, suggesting to some that colonization is not a one-way process.

Aside from the profound shock Smith experienced at their meeting, Pocahontas may most have unsettled him by the masque she produced for him. Masques were a dramatic form Smith was familiar with in England; however, he was entirely unprepared for being the honored guest at a forest version of what he couldn't help but think of as a courtly kind of affair. His description of the event makes his shock at its inception clear. He regained his poise soon enough, as his account also shows, when he realized that although the costuming and routine were unfamiliar to him, the form was not. He would later refer to this performance as "A Virginia Mask," a term that was more usually spelled "masque" in court circles in his day.

It seems that in the late summer of 1608, not too many months before he would return to England, Smith was given a mission: Captain Christopher Newport, the recently arrived head of the mission, ordered Smith to go to Werowocomoco to arrange for Wahunsenacawh's coronation. The assignment annoyed Smith because he believed that Newport was all too liberal in his treatment of the Native people, undoing all his, Smith's, efforts. He felt, strongly, that the idea of crowning Powhatan "Great King and Emperor of the Powhatan" would only serve to give the priest-king grandiose notions of himself and the prominence of his people in world affairs. This, Smith fumed, could only lead to disadvantage in trade for the colonists as well as to increased threat of violence when the Powhatans' trade demands were not met. Smith was of the iron-fist-in-velvet-glove school of diplomacy, and had little regard for more convivial modes.

He addressed a number of his complaints to the treasurer and council of Virginia; they were recorded as being "from Captaine Smith, then President in Virginia." Specifically commenting on matters pertaining to Wahunsenacawh, he wrote, "For the Coronation of Powhatan, by whose advice you sent him such presents, I know not;

but this give me leave to tell you, I feare they will be the confusion of us all ere we heare from you again."[2]

The fact that his office of president of the Jamestown enterprise was lost with Newport's arrival might have had some influence on the spin he put on his remarks. He was perhaps unduly concerned with his image, and was careful to document his competence, enlisting the testimony of members of the company well disposed toward him to back up his claims. Little good his trouble did his career. It hit a downward trajectory when, seriously wounded, he returned to England, never to quite recover the same status of adventurer with the missions to match that he had enjoyed up until he met with, and betrayed, the Powhatan Great Council.

Unhappily based at Werowocomoco, Smith had some time to wait for Newport, who had gone upriver on an exploration trip required by executives in England, to make his appearance. Perhaps to help make the time pass more congenially for her charge, Pocahontas saw to the well-being of the English contingent, arranging for their entertainment. She had the five honored English guests assemble in a meadow near the town, a place Smith described as "a fayre plaine field." They were given seats on mats near the center fire, itself an honor. Undoubtedly the fire and the "fayre plaine field" had been appropriately blessed before the proceedings because most such activities were—as they still are. The blessings are normally accompanied by honoring fire manito with copious amounts of cured and blessed tobacco; the honoring ends with a humble request for manito's beneficent overseeing of the procedures. Likewise the ground would have been honored, earth manito's blessing and beneficence requested. Some pipe smoking would have been part of the ceremony, to honor the air manito. As for the fourth element, water manito, participants would bathe before the event. "Going to water," ritual dipping and praying accompanied by the offering of blessed tobacco in the running water of a river or a stream, is still an obligatory part of these occasions.

Included at the event Smith described were a company of Powhatan people, and the English settled among them. Soon they heard quite a commotion: as Smith described it there came "amongst the woods . . . such a hydeous noise and shreeking" that the startled Englishmen reached for their weapons and made ready to defend their lives. They also grabbed two or three old men, "supposing Powhatan with all his power was come to surprise them."

Quick to respond—however startled by their rush to violence she may have been—Pocahontas stood clear of the crowd, put one hand over her heart and the other raised to the sky and commanded John Smith "to kill her if any hurt was intended." Of course, the other Indians did what they could to reassure the volatile visitors, and so the guests put down their weapons and resumed their places on their sitting mats. Smith details the event:

> . . . thirtie young women came naked out of the woods, onely covered behind and before with a few greene leaves, their bodies all painted, some of one colour, some of another, but all differing, their leader had a fayre payre of Bucks hornes on her head, and an Otter's skine at her girtle, and another at her arme, a quiver of arrowes at her backe, a bow and arrowes in her hand; the next had in her hand a Sword, another a club, another a pot-sticke; all horned alike: the rest everyone with their severall devises. These fiends with most hellish shouts and cryes, rushing from among the trees cast themselves in a ring around the fire, singing and dauncing with most excellent ill varietie, oft falling into their infernal passions, and solemnly again to sing and daunce; having spent neare an houre in this mascarado, as they entered in like manner they departed.

Having reaccommodated themselves, they solemnly invited him [Smith] to their lodgings, where he was no sooner within the house, but all these Nymphes more tormented him then ever, with crowding, pressing, and hanging about him, most tediously crying, Love you not me? Love you not me?

This salutation ending, the feast was set, consisting of all the Salvage dainties they could devise: some attending, others singing and dauncing about him.

These bewildering activities continued, he said, "until in the end, lighting the way with firebrands instead of Torches, they conducted him to his lodging."[3]

In his recorded account, Smith seems to have missed the point of the "mask" entirely. He sees the young women, "Nymphs," whose attempts to have sex with him were part of a ceremony in which he was invited to enter deep communication with the manito. These spirit beings, in the way of tribal religion, were present *in* the dancers, whose bodies were, ceremonially speaking, vases, containers, for those presences, and continued so for the duration of the ceremony, which lasted until the men were escorted by firelight to their beds. Obtusely, Smith viewed this stunning invitation as even greater torment than the astoundingly European pagan-reminiscent ceremony Pocahontas and her ladies performed in his honor. His was a case of "many are called, but few comprehend the summons." He was at least marginally aware of its portents, though. He characterizes the participants as "fiends with most hellish shouts and cryes," showing that he knows the universe of discourse in which they are operating. As a Christian, and furthermore a man who prided himself on his "reason" and "modern" outlook, he could not get beyond his fear of "devilish" forces, which was how he had been raised to view those same forces still recognized in his own culture. Nor was he unbelieving: he recognized the arcane or supernatural nature of the dancers, their attire, and their movements. His very repudiation of them, his intense denial that he found them in any way attractive, speaks volumes about his awareness of what had transpired, despite his attempt to frame his personal experience in familiar Christian terms.

Over the months of Smith's sojourn in Jamestown, Pocahontas's visits shifted somewhat in their intent as she grew older; however, she continued to grace the fort with her presence regularly until John Smith went home, leaving Pocahontas with the misperception that he had died. Wounded in a freak accident that suggests to some that the manito didn't take rejection gracefully, Smith was forced to head to England, where he could recover. This was much against his own will: he planned to settle outside the realm of the Virginia Company's interest, having secured a site from Parahunt, a Nansemond Indian friend. Evidently no one told Pocahontas that Smith, her English spiritual charge, had gone home; she seems to have believed him dead. Some writers suggest that John Rolfe confirmed that idea while he was courting her, and that she agreed to the marriage out of displaced longing for her lost lover. Should there be any relationship between the deceit and her choice of Rolfe as a substitute, the connection would be more about the charge she held from the manito and less about her romantic impulses. She was, after all, a woman of the Dream-Vision People, and to her, as to her people, the Dream-Vision, not adolescent yearning, was—as it remains—the primary motivator.

Again, an interpretation of this sort depends on a traditional Anglo-American point of view, although it can be noted that she may indeed have decided that since her first English spirit charge, Nantaquod, was dead, she would have to find another intelligence source. While this reading comes to approximately the same conclusion as the more traditionally feminine one—in American versions, that is—its import is quite different. Seen as an account of the maneuverings of a Native espionage agent in a hostile, alien environment, a Nativistic interpretation goes a long way toward reconciling a number of the seeming contradictions in her life.

As the official emissary of the Council at Werowocomoco—later relocated to Orapaks, which stood at the juncture of the Chickahominy and Pamunkey Rivers—Pocahontas often entered the fort leading dozens of her country people bearing large quantities of corn, game, and other edibles. However, her role changed as relations between the *tsenacommacah* and the interlopers worsened.

Although commanded by Powhatan to sever her relations with the colonists, Pocahontas remained sympathetic to them, helping them in countless ways whenever she encountered

them in Powhatan territory. Though her task grew considerably more difficult as the struggle between her "father" and
the colonists moved toward its climax and she no longer
openly aided the colonists or served as emissary, nevertheless
on at least two occasions in 1609 she intervened to save English lives.[4]

While "Powhatan," who the English and subsequently the historians
and biographers believed was an absolute monarch as King James was,
was the paramount or principal chief of the *tsenacommacah,* it is unlikely that he ordered people about. Wahunsenacawh, known to the
English as "Powhatan," was the greatest *mamanantowick,* shaman-priest
and political leader, of the *tsenacommacah,* the Powhatan Alliance,
though the Grand Council made decisions in a way that the English
system resembles today. That said, the fact was that the situation was
increasingly conflicted; battles broke out, and Smith, as president of
the James Fort council, made improvements to the fort's fortifications,
protecting it from attacks. Meanwhile, the first real settlers were arriving, and Englishwomen were added to the contingent.

Their own efforts were poorly rewarded, the Englishmen for the
most part unable to provide enough game, fish, or clams to supply
their needs, and the escalating violence left the Indians less than willing to provide for those whose presence meant their own destruction.

In the winter of 1609, at the invitation of the *matchacómoco,* the
Grand Council, Smith led a company of men to Werowocomoco, the
main food storage center of the *tsenacommacah.* The *tsenacommacah* was
no longer sending CARE packages, and the English, who stub-
bornly remained in the area despite Smith's assurance to his "father"
Powhatan, were hungry. Their earlier attempts to trade for food had
ended on one occasion in bloodshed, and on others with insufficient
supplies or none at all. The struggle got bloody when Smith led a
party to Nansemond country, taking two barges he planned to fill
with about four hundred bushels of maize that had been promised the
English. The Nansemonds were part of the Powhatan Alliance, so

when Smith landed at their main town he was told that due to the war, the deal was off. Furious, Smith and his men set fire to the first house they saw, and the Nansemonds yielded. They gave him about half of the agreed-upon load, soon enough exhausted.

About the time starvation was looming once again, Smith received word that the Powhatan Alliance would provide the colony with generous provisions if Smith would provide the transportation. Smith agreed, naming Werowocomoco as the location where his men would load the provisions. They met, and after some futile negotiations over the Indians' right to bear arms, neither he nor Smith giving an inch, Wahunsenacawh departed the negotiations, convinced he was in danger from English weapons.

Seeming to reappraise his position, Wahunsenacawh sent word to Smith that he would supply the food after all, requiring only that Smith himself remain to oversee the loading. Smith agreed, settling down to wait until the corn was ready for loading. In the night Pocahontas appeared and told Smith—perhaps all the Englishmen on the job there—that Wahunsenacawh intended to kill them all. Seeing that his negotiations had not gone well, and considering that he had received other hints of treachery, Smith made ready to once again nobly defend himself and his men on his way to meet with Wahunsenacawh, as arranged. When the Indians appeared the next day with food that Wahunsenacawh had sent to fortify them for the day's work, Smith had one of the food bearers taste each dish before allowing his men to eat. He feared poisoning, because one of the Indian informants downstream had suggested that poison would be the old priest-king's chosen method of attack. However, the meal was unpoisoned, the loading went well, and the English hightailed it out of Powhatan country as soon as they could. As Smith would retell it in later years, he had once again been saved by the beauteous and loving Indian princess who betrayed her father to save the English and the life of her beloved John Smith.

Once again, Pocahontas's behavior is subject to various interpretations. The question of why she betrayed her people, or whether she in fact did so, is raised only because of Smith's account of the event.

Smith, however, was writing years after the fact. Years after Pocahontas's death he was not above name-dropping. Still angling to gain berth on promising adventures, he needed to curry favor with the movers and shakers in England. Since Pocahontas, by then Lady Rebecca, had been wined and dined by the rich and powerful, and, though now dead, had made quite an impression on the powers he was courting, this tale could have been made of whole cloth. However that may be, he made much of his close relationship with the Indian princess, and while it never got him back to Nansemond country or his dreamed-of plantation, Nonesuch, it seemed to have given him credence with historians from his time to this.

On the other hand, if the events he described did indeed occur, the why of her actions is not all that clear. Interpreting and judging her actions depends on whether the story is told within a modern English-American context or within an Algonquin one. If one takes the manito into account—as one must when considering Pocahontas outside the confines of English business and religion politics—her actions in terms of Captain John Smith take on a cast that greatly differs from the accepted Anglo-American one.

For three years after Smith left her country, Pocahontas was not seen by the English. She was doing whatever she needed to finish her training in the ways of the *midéwewin,* of the manito way, getting her chops as a fully adult woman of high spiritual and clan status in her own right.

The abduction of Pocahontas by Samuel Argall in 1612 is revealing in this regard. It illuminates the strange yet eerily familiar conflation of piety and greed while it raises ever more perplexing questions about Pocahontas and her motives in boarding, seemingly oblivious, an English boat headed, as she had to know, for James Fort. A consideration of that pivotal event casts light on an entire sequence of events: the paradoxical possibilities set in motion by her "abduction" in the end clarify her goals, her roles, and the significance of her short life. Again, it is clear that the significance of that life rests largely on her role in the manito-directed drama, events sourced in the implicate order of her world and how that order affects the one moderns think of as reality. After all, paradox is a major identifying characteristic of implicate-order events, particularly so when they merge with more mundane processes.

Simply put, the implicate order, an idea fostered by the physicist David Bohm, is the subreality of energies before they take on shape or form as we recognize them. In contrasts with what Bohm termed *the explicate order,* in which events, people, objects, even planets and galaxies, take on the guise human brains recognize and consciously interact with. In Algonquin terms, the implicate order Bohm details is hugely analogous to the world of the manito, the mystery that can be both force and being, wave-form or particle, identity or surmise. The explicate order, on the other hand, is rigid "reality," which defines objects large and small as fixed in nature or essential being. A rock, for instance, is a rock—igneous, metamorphic, silicate; broken, whole, polished, rough—a mineral constructed of measurable, definable molecules that hold to a definable pattern of organization. In the *powa* world, the world of Dream-Vision, physical measurement becomes irrelevant. As all is in constant flux, any events or objects can change from one "form" to another, one "state" to another, and one "significance" to another, depending on a constellation of probabilities. In this order, the manito, the "mystery"—great or small—is the deciding force, the intelligent observer that determines what shape or event will arise, and when, in the explicate end of things.

There is little question, given the testimony of Englishmen who participated in that kidnapping, that more than mere hostage taking was planned. Had not Argall himself likely heard the passionate sermon delivered to men about to embark for Virginia Company's stronghold across the Atlantic? Frances Mossiker, in *Pocahontas: The Life and the Legend,* observes:

> That the suggestion of kidnapping in connection with the conversion of the heathen was heard even from the pulpit: in May of 1609, the Reverend William Symonds reminded his Virginia-bound congregation that "a captive girl brought Naman to the Prophet. A captive woman was the means of

progressive movements of the 1960s and beyond. Histories of Native people, stored and recounted by Natives who were there, testify quite otherwise. While in the long run, or so it seems now, the Indians lost the struggle, it is incontestable that struggle they did. Not only in obvious modes such as overt warfare, but in less dramatic ways: they retained their understanding of how the world goes, of what constitutes reality, and of how to live a life that is in harmony within the greater mystery of all-that-is. This particularly Native kind of resisting by retaining old mores and old ways persists; over the past several decades it has been slowly recovering its strength.

This observation leads to further consideration: if they did not just vanish into the forest, never to be seen or heard from again (a favorite, if sad, belief among Americans), what did they do? How did they interact with the newcomers? What skills, understandings, abilities, and maneuvers of resistance and survival did they employ to counter and adjust to the new circumstances in which they became situated?

In the face of such considerations, the possibility that Taylor's espionage scenario might extend to other Native women whose lives paralleled Pocahontas's life seemed well worth exploring. In addition to Pocahontas, Malinalli, the woman enslaved by the Aztecs who is known in Mexican history as *La Malinche,* and Sacagewea, a Shoshone woman, taken as hostage-slave by the Hidatsa in her childhood, are excellent candidates. Not only do their lives offer a wide difference in time and geographical location, but despite these differences a large number of factors coincide in their biographies. Each was taken hostage by an enemy society; each enjoyed the attention of powerful European invaders; and each has become a major historical and popular figure.

Native women played a critical role in the defense and protection of their own communities. It is well to remember, in this regard, that in their times, as in ours, the idea of "Indian" as a monolithic group of humans is more apparent to outsiders than to people within the various language and culture groups. The histories of Malinalli and Sacagewea, who came from early sixteenth-century *Mexica* (Mexico, now) and the eighteenth-century Rocky Mountain region of North America, respectively, help clarify the role that Pocahontas, historically positioned between the two, played.

The histories and actions of all three women possess a sufficient number of parallels to indicate that their lives were of a pattern, and as such point out new avenues of inquiry into the history of relations between Native and foreigner in earlier centuries. Helpfully, the other two, as women, were abducted by Indian communities hostile to their own. This fact helps clarify quite a bit of both Pocahontas's otherwise puzzling behavior and the competing possibilities that shed light on her life. Unlike Malinalli and Sacagewea, Pocahontas was not enslaved in childhood but was a fully functioning, high-status young woman among her own people until she was fully grown. It is possible that her lifetime at Werowocomoco as some kind of "adjunct" to the Powhatan, the priest-king Wahunsenacawh, was in itself a local brand of enslavement. Such is not out of the question. The Oral Tradition is rich in lore about the taking and holding of women who possess great spiritual or magical power, which is harnessed to the use and benefit of some great sorcerer. The Algonquin tradition itself is based on such a situation, which lends some credence to the possibility, although such cases do not include children, to the best of my knowledge.

In the traditions of the various Native Nations of the Americas, the role played by women taken hostage by supernatural entities is conspicuous. In those narratives in which a particular woman plays a significant role, she develops an ongoing relation with the supernatural world she is taken to. In such stories she usually becomes wife of the abductor, himself a major holy person, and, in taking on her duties as his mother's charge, fails at a variety of tasks. These failures lead to the woman's gaining supernatural status of her own, and as a consequence she can return to her people and provide them with various kinds of sacred knowledge.

Malinalli (*Malinal*) served as adviser and strategist to the Spanish conquistador Captain Hernán Cortés in the sixteenth-century conquest of Mexico, and Sacagewea (usually misspelled "Sacajewea"), the young mother, accompanied Meriwether Lewis and William Clark on their cross-continental journey of exploration in the early nineteenth century at the behest of the man who was then president of the United States, Thomas Jefferson.

Malinalli—*La Doña Marina,* as the young Spanish soldier Bernal Diaz del Castillo, whose journals chronicle his company's foray into the New World from 1519 (*ce atl* according to the Aztec calendar), unfailingly called her—was the earliest of these fascinating, powerful women. She is best known to those familiar with Mexican history as *La Malinche;* the title means Head (Woman), or, as often translated, "Chief" (feminine). Her personal name, Malinalli, means "penance grass"; the name might allude to a religious practice among Aztecs and other Native Nations in Meso-America that made use of this plant. The practice involved making offerings of their own to their ever hungry gods. The sacrificer might pierce tongue or penis with the razor sharp blades of penance grass, drawing them through the flesh and leaving them in for a time. The blood was caught in a cup and offered to one or another supernatural being.

While actually part of a complex religious system entirely alien to Christianity, this practice has been likened to Christian cultural concepts of penance; thus the Spanish/English translation of the Nahuatl *Malinalli.* The word might better be translated "cactus spine" or "cactus thorn." Personal piercing as an offering of self to some supernatural being, force, or harmonic (depending on the Native American group, its culture, and the context in which the sacrifice occurred) was a widespread practice throughout the Americas. It is still practiced by traditionals in ceremonies that date back to "time immemorial."

As history demonstrates, Malinalli was as well named as was Pocahontas: her name reflected her nature and the part she played in what I have come to view as the sacred history of the world. Malinalli's work with the Europeans as guide, diplomat, and translator tipped the balance of power away from the powerful Aztecs, who had enslaved her and thousands of others, and toward the Spaniards, in whom she saw her liberation and that of her own people. In that all but miraculous way, she played the role, society-wide, that her plant cousin *Malinalli* played for individuals and smaller groups. In murals of that time she is pictured as a very large figure, looming over her smaller Aztec and Spanish companions. The American historian William Brandon tells us of her brilliance at military strategy, as well as chronicling her linguistic skill:

> She spoke Nahuatl as a birthright tongue, and in Tobasco she had learned the border Maya dialect known as Chontal, and also, possibly from merchants visiting the household where she lived, Maya proper, as spoken throughout Yucatán. She could talk to Aguilár in Yucatecan Maya, and Aguilár could then translate into Castilian for Cortés.[7]

The foreign invaders had weapons that were slightly superior— "but not that much superior," Brandon observes; their numbers were around four hundred—to those of the maybe 10 million Native peoples in the region. Each Mexican city could call up rank upon rank of trained and ready soldiers, so the tiny company under Cortés's command was so puny as to be laughable. The steel breastplates the Spaniards wore were climatically inferior to the quilted cotton armor of the Aztecs, and they were attempting to fight in a humid region of the world that was about as strange to them as Mars or Venus would be to modern armies. Yet these bearded strangers, undermanned, outgunned, and easily outmaneuvered, gained control of a realm four or five times the size of Spain, more than twice as populous, and considerably more wealthy.

As in the later conquest of Virginia and the rest of the northern Atlantic seaboard, there were many factors in play in this unimaginable triumph. One that to the Spaniards seemed incontestable was the intelligence of the "Lady" who guided them.

"The important point," Brandon instructs us,

is that throughout the first march on Mexico, after they were joined by Malinalli, the Spanish were forced to fight in only one instance—where only their immensely superior tactics saved their lives. Otherwise, the road of their first penetration into the country—the perilous interval while they were still without important allies and could have been wiped out a dozen times over—was paved by a string of diplomatic victories as remarkable as so many straight passes at dice.[8]

Like Pocahontas, Malinalli would become an indigenous American version of Eve, only in her case it was Chicano protest rhetoric in the 1960s and '70s rather than that of American Indian Movement members or that of the activist community at large. Known among this community as "*La Chingada*" (which translates as the sexual act in its lewd sense, feminine case), Malinalli was blamed for the European conquest of the Native peoples of what would become Mexico and areas farther north and south of its present borders. Following their Chicano brothers' lead, American Indian activists began to charge Pocahontas with similar accusations. She would become the most famous female "Apple" (red on the outside, white on the inside) of the era. Unlike Pocahontas, *La Malinche* has been a familiar figure to me since early childhood. I was raised in New Mexico, not too long ago part of Mexico, and my early ideas of conquest and colonial history included *conquistadores,* and the language of the conqueror was Spanish. Pocahontas, on the other hand, was, to a New Mexican like me, from a basically alien part of the world and part of its basically alien history.

Like Pocahontas, Malinalli was safely married off to a suitable European, becoming the Lady (*La Doña*) Marina, good daughter of the Spanish crown and the Roman church. Like Pocahontas, Malinalli was mother to a son, fathered by the famed (or notorious) conquistador Hernán Cortés.

The ways of the conquering foreigners from across the sea are strange, and dully predictable: if it's female, use her strength, ability, and charm to achieve your aim, then marry her off to a mild, non-threatening male under your control. It is helpful if she is seen as "comely," as was Pocahontas, or "good looking . . . and without embarrassment" as the soldier Diaz del Castillo wrote of Malinalli. Even more helpful, when touting her as a model of new Nativism, is to emphasize that she is safely civilized and Christianized. Dressed in a manner that befits a lady, and safely under the control of husband and church, she can be held up to young Indians with pointed pride. You might write of them as Bishop Landa did of Lady Marina: ". . . and thus God provided Cortés with good and faithful interpreters by means of whom he came to have intimate knowledge of the affairs of Mexico, of which Marina knew much . . . ," or so Brandon quotes the good bishop's translated remarks. Made to disappear under the skirts of decent Christendom, these indigenes can safely be dismissed from history. Should they resurface, it's best to politicize them by categorizing them as collaborators. Thus, reconstructed from friend to foe, they can once again be safely erased from common memory.

On the whole, Sacagewea has fared better, historically speaking, than her foresisters; at least she hasn't been particularly vilified or banished from memory. In fact, the most recent dollar coin is graced with her name and image, which was modeled by a young Shoshone woman from the Wind River community. However, much of Sacagewea's story has been distorted and mythologized, as have the stories of Pocahontas and Malinalli.

In the case of Sacagewea, though, the mythologization and distortion were more an embarrassment of riches, so to speak, than diminution: a leading suffragette saw in Sacagewea a noble symbol of the strength of the female sex, believing that Sacagewea had led the famed 1805–1806 Lewis and Clark expedition. Geared to locate interior water pathways through the continent from St. Louis to the Pacific, it succeeded in opening the west to American settlement. It was the first expedition to find its way across the country. Earlier attempts had been stymied by Native resistance to white passage, as earlier attempts to penetrate into the Yucatán met with both fierce resistance and devastating illness until Cortés. But the expedition was under the guide of a rough French trapper named Toussaint Charbonneau. Sacagewea, then around eighteen years old, did not serve as its guide. She carried her small child on her back on its cradle board until the board was lost in a sudden flood. Then she carried the boy in her arms or held securely in her shawl. Sacagewea accompanied the men.

> Enemies from another tribe kidnapped Sacajewea who was a Shoshone Indian. These Indians sold her to a French-Canadian trader named Toussaint Charbonneau. Charbonneau was hired as an interpreter when they joined the Lewis and Clark expedition as it passed up the Missouri River in 1804. However, it was Sacajewea who proved useful when she saved them from harm from the Shoshone Indians. She was also able to get food and supplies that the travelers needed from her relatives she met when crossing the Continental Divide. She was an extraordinary woman and there are many memorials dedicated to her, the most famous one located in Washington Park, Portland, Oregon.[9]

There are more statues of Sacagewea in the United States than of any other woman; she is so renowned that there is a mountain peak named for her, Sacajewea Peak, located in the Bridger Mountain Range in Montana.

One imagines that she did various chores, the main one ensuring that Charbonneau would serve as guide. She did exert her connections on a couple of occasions: once when they entered Hidatsa country, and the other when they were in the Sierra in the lands of her people, the western Shoshones.

After their return to St. Louis and Charbonneau's other women, Sacagewea fled. Charbonneau had beaten her dreadfully. As soon as she could move, although barely able to crawl, she made her way out of his reach. When she recovered she used the presidential medallion she, along with the other participants in the expedition, had received from Jefferson to make her way all over the West. She wanted to see how her people fared, she said. Eventually she settled among the Commanche, where she married and raised several children. When her husband died, Sacagewea made her way back to her home, among the Shoshone.

Although Sacagewea was never seen as aristocracy or royalty, being a figure from early nineteenth-century America, she bore a lifetime honor bestowed upon her by President Jefferson, one she shared with several other prominent Americans—a presidential medallion. This piece was extremely valuable in prestige and as a tool. Enabling the holder to free transport by any public means, as well as to hospitality such as bed and board, it freed Sacagewea to travel around Indian Country, satisfying her desire to discover how fared the other Native peoples of the then American frontier. By the standards of the time, this was major official acknowledgment. She used it in her later years after she returned to her original home to aid her Shoshone people in wangling a decent land settlement from the federal bureaucracy. She succeeded, despite the attempts of bureaucracy, politicians, military installations, and traders seeing progress and attendant wealth in the rich vastnesses of the Louisiana Purchase, where whiskey and gun running accompanied land development and military supply scams much as drugs and armament industries do today.

Like Pocahontas and Malinalli, Sacagewea had another name, or, more properly, title. During the many years she lived among the Commanche she was known as Porivo, which in English means "head" or "chief" (feminine). In this respect her story echoes that of Malinalli, as does her life as captive of a Native Nation inimical to her own. Sacagewea's life also paralleled the lives of Pocahontas and Malinalli in that each woman served as intermediary between the supernaturals and incoming systems that differed greatly from those preceding them. And each woman's legacy has far outweighed her brief moment upon the stage of New World history.

In the histories of these three women, in a world where female presence is usually seen as an adjunct to male activity, we see something out of the ordinary going on. From earliest contact on the mainland through the ensuing centuries of contact, exploration, settlement, and conquest that culminated in the U.S. Army's massacre of Black Kettle's small band at Wounded Knee in 1891, few names of women come down to us. These three have all been perceived as traitors to indigenous people; each was abducted, each made history, and each left a legacy that informs this century as much as, if not more than, it impacted their own. Each occupied a leadership position among her own people, and acted as an agent of change, bridging worlds so that eventually harmony might ensue.

SACRED SPIES

Among the indigenous women who appear in modern histories perhaps these three, Pocahontas, Malinalli, and Sacagewea, make the best candidates. The earliest was Malinalli, Cortés's adviser, strategist, secretary of state, lover, and the mother of his son. Second, and subject of this study, was Pocahontas, the spiritual mother of John Smith and the Virginia Company and the source of their tenure on this Turtle Island. She was also wife of the planter John Rolfe and inadvertent founding matron of the American plantation system via its first exportable crop, tobacco. The third and most recent was Sacagewea, sometimes guide and savior of the expedition that opened the entirety of the American continent south of Canada and north of Mexico to the government and the people of the United States.

If these women were spies, whom could they have been working for? To begin to respond to that query, we must enter into a long digression into the workings of the Oral Tradition as it interacts with the belief and worldview of tribal people. Understanding Pocahontas as a manito woman in the making, how "magic" and ritual combine to create effects in real time, is in order. Her story is about the meeting, the interface, of the implicate and explicate orders. It is about how a standing wave forms into a manifest phenomenon—that is, how the *tsenacommacah* morphs into the modern Virginia we recognize, while it remains the land of the manito, the Dream-Vision, the place where potential can become actuality.

In the case of Pocahontas, an exploration of some of the stories from the Oral Tradition that have abduction as their theme is in order. Since our subject is Pocahontas, a figure who is assuming mythic status, the traditions concerning "sacred" women will help establish Pocahontas's identity and role in the implicate order. These figures are identified as "sacred" because of their pivotal role in relations between supernatural and human beings. When their role is fulfilled they attain a status of supernatural themselves. In many of the stories about such figures, abduction plays a major role in the unfolding of sacred events.

The Oral Tradition is commonly believed to be a body of lore handed down generation by generation. It can include just about anything, but usually refers to stories, songs, ceremonies, herbology, hunting and fishing lore, and other such matters of special interest. It is assumed, and frequently stated, that the Native Nations did not have writing systems, and thus depended on the Oral Tradition to store and transmit information.

While it is true that the Oral Tradition stores and transmits knowledge, it is not true that it does so without means of ascertaining the accuracy or provenance of the information, its sources, and the authority upon which it rests, or that these means are purely oral themselves, as when an elder checks the practice or narration of a younger person. A cursory examination of rocks all around us in the United States alone give the lie to this idea. There are texts inscribed in stone, and these texts, referred to as *petroglyphs* or *pictographs,* serve both as storage and transmission. Additionally, every Native society has more portable means of storing and transmitting bodies of knowledge. These include pottery, blanket, basket, and bark-implement designs, totem poles, masks, apparel, mats, storage containers of wood or other materials; the list goes on. Architecture is a major information-storage mode, as are dance, music, and geographical features. Some of the old ones used to say that the land was our book, and, given the way they "saw" and "read" the living teacher—a kind of "talking book" that is self-aware, intelligent—their meaning becomes evident. How they do this is about stories. Trying to convey the sense of this interaction analytically is more confusing than helpful, which is why the Oral Tradition takes the form it does: narrative conjoined to ceremony. The stories that follow are not explicitly about Pocahontas; rather, they are stories that enable me to understand how she might have lived, the kind of mental universe she inhabited. In telling them I hope that the way in which the two worlds, implicate and explicate, interact to make each day what it is becomes more clear, and that clarity leads to a fuller grasp of how a particular human being (rarely, but sometimes) becomes a manito.

Among my people, there is a group known as the *shiwanna*. These beings are Rain or Rain Cloud People, and they come to bring rain. Before one thinks "Oh, animism," or "A primitive [or primal] explanation of a natural event," consider an experience I enjoyed. Around 1972, I took my mother to Feast Day at Santa Ana Pueblo for her birthday, which is July 26. New Mexico had been experiencing a drought for nearly fifty years, and that July was no exception. I had listened to the weather forecasts on the radio, and of course on the local television news, and there was no rain predicted because there was no weather front to be seen for a thousand miles around. That day dawned hot and dry. The sun was, as usual there in July, intolerable, the heat exhausting. Braving the heat, we drove north from Albuquerque, arriving at the pueblo around noon. We had to wait for the dance to resume, but spent our time visiting some friends who lived at the pueblo. In time the dozen or so dancers—women, men, and two or three children—emerged from the kiva, and the dance began again. They had been dancing since early morning, and would continue until near sunset, with only the noonday break. The dancers' bare feet, moving in soft, gentle rhythm to the soft beating of the small drums the drummers carried, raised a small cloud of dust around the dance ground. The sun, the heat, were brutal. Pouring sweat, the dancers kept on, faces calm, movements deliberate, as Pueblo actions generally are. After about an hour of concentrated dancing, the wind kicked up. Swirling dust devils rose beyond the village, and suddenly we caught a whiff of that unique, exciting New Mexican summer scent of rain on the cooling wind. Within a few moments the rain began, and the dancers' dusty faces were soon streaked with mud, their calm expressions lit by triumphant grins. The *shiwanna* answered the call of the dancers, though rain was not in the offing, scientifically speaking. That evening's news didn't mention any sudden weather fronts for that day, although, driving up to the pueblo that morning, I had seen some Cloud People, *shiwanna,* rising up from the Sangre de Cristos; ground clouds they were, just like the tradition says.

That is because the *shiwanna* are implicate-order beings, while weather forecasting is an explicate-order affair. The dance, and the events and preparations surrounding it, including its annual date and position on the Pueblo calendar, taken together serve as bridges between the two orders of reality. Dances, as the Pueblo call what might more generally be termed "ceremonies," are not always rain bearing, any more than weather forecasts are always spot-on. The reasons why the rain people don't come when called are many, chiefly because they are not projections of human need or wish but independent beings who make their own decisions, regardless of mortal blandishments. Which is why Old Lodgeskins, in the 1970 film *Little Big Man,* starring Dustin Hoffman, wryly remarks: "Sometimes the magic works, and sometimes it doesn't."

In the life of Pocahontas there are a number of events that parallel both the advent of the rain people at Santa Ana in 1972 and the event that left Old Lodgeskins, though ceremonially prepared to die, having had foreknowledge of his death's imminence via spirit message. An example of the first is the Feast of Nikomis, when Smith was remade into Nantaquod. A certain kind of intelligence-energy—a standing wave from the implicit order, if you will—was brought to bear on the people gathered in the *quioccosan,* the sacred grand house, or temple. This presence led to the determination that Smith was to be remade, not killed, and the bearer of the message was Pocahontas, who became responsible for Smith/Nantaquod's further existence. As for the wonderful words of Old Lodgeskins, "Sometimes the magic works, and sometimes it doesn't," all one has to do is consider how the remaking of John Smith turned out to take his point.

CIRCLES WITHIN CIRCLES EVER INTERTWINING

John (Fire) Lame Deer, late medicine man of the Lakota, devotes his memoir, *Lame Deer: Seeker of Visions,* to clarifying and wittily commenting on the Lakota tradition as he practiced it. In one chapter he tells about some of those "natural beings" and how they are seen by shamans such as Pocahontas. He begins with some explanatory comments: "Nothing is so small or unimportant but it has a spirit given to it by *Wakan Tanka* (Creator)," he instructs.

> The gods are separate beings, but they are all united in *Wankan.* . . . You can't explain it except by going back to the "circles within circles" idea, the spirit splitting itself up into stones, trees, tiny insects even, making them all *wakan* by [Creator's] ever-presence.

He goes on to name some of these spirit beings, spirit-sharing beings, spirit-informed beings:

> Tunkan, the stone spirit; Wakinyan, the thunder spirit; Takuskanska, the moving spirit; Unktehi, the water spirit— they are all *wakan:* mysterious, wonderful, incomprehensible, holy. They are all part of the Great Mystery.[10]

Lame Deer might have added plant people to this list. Plants such as corn, beans, squash, and tobacco, along with somewhat better known "sacred" plants such as peyote, *ayuasca,* cocoa, jimsonweed, henbane, and nightshade, about which we will have more to say in chapter 4, "*Apook /* The Esteemed Weed," are also part of that vastness referred variously to as Creator, The Great Spirit, All-That-Is, and the Great Mystery. They too are active, and sometimes proactive, members of the larger community of which human beings are but a small part.

At their Feast Day dances, the people of San Filipe Pueblo in New Mexico include plant and animal people along with human and supernatural beings in the ceremony, making a whole that is at once nourishing to the spirit and evocative of the greater Being of the Earth herself. The dancers eerily make the beings they are "dancing" as vitally real as any of the other figures being danced, forming a field of meaning that goes well beyond the ordinary or mundane. These are not theatrical performances; they are instead the source from which theater derives, in primary and pure form, even though they are not taking place in ancient Greece but in contemporary Native America. The "mask," as John Smith termed it, that Pocahontas held to honor him was a similar ceremony. The young women, by Smith's testimony antlered, white deerskin draping their shoulders and lower torsos, must have appeared much as the Feast Day dancers. There are Deer Dances held at many of the pueblos in the autumn, around deer-hunting time, and the men who dance the part of the deer soon become indistinguishable from deer. It is not only that their movements are perfectly mimed; it is that the Deer Spirit has entered them, and by that presence has agreed to let the hunt go on. The maidens "dauncing" with Pocahontas became the deer women they invoked; thus Smith decried them as "devilish" and "hellish." His eyes could not deny the manito presence, although his culture accounted it evil rather than holy.

It is the tradition that holds the memories, the significances of the ceremonies. But the words are meaningless absent actual experience. It's as if someone from another star system were trying to write a description of pregnancy and childbirth for people who don't bear their offspring. Imagine how outlandish to the readers would seem the idea that a seed gets planted in a body, grows and grows, making the seeded person swell and swell, and then out pops a fully formed being. The words could not do justice to the event, but would misinform and mystify. But having seen a birth, having gone through pregnancy and labor, and then to read descriptions of the process not only makes comprehension of the material read possible, but also empowers informed response.

Similarly, the events in the life of a sacred spy such as Pocahontas may seem puzzling if not incomprehensible to those who have neither participated in nor witnessed similar events. Readers who have themselves participated in or witnessed such events will use those same accounts as reminders of their experiences.

The tradition is not confined to the Oral. That is, it does not depend wholly on verbal transmission from one person or generation to another. The People have always used glyphs to record information. These glyphs appear on rocks, bark, leather, stone stelae, papyrus and linen writing surfaces, paintings, cave walls, pottery, weavings, bark bags, woven baskets, jewelry, tattoos, and a multitude of other such surfaces. It's not that the Oral Tradition is actually "oral" so much as that one of its most familiar modes—to outsiders and scholars—is the "as told to" genre. An informed spectator-participant at traditional Indian dances is able to "read" the dance just as a trained astronomer can "read" the stars or a trained auto mechanic can "read" the pings, coughs, or sputters of a car engine.

Equally, a trained medicine woman can "read" the land—the trees, the rocks, the winds, the geological formations, the movements of clouds, streams, ocean tides, stars, the flights of birds—as accurately as a highly literate professional in the health professions can read a medical text or an attorney can read a legal brief. People such as Pocahontas are professionals of the former kind. She could "read" the time, the weather, the eddies and currents of human interactions on a community scale, the teachings of the tobacco in its cured leaves, its growing plants, its smoke patterns, and so on. More accurately put, she could "hear" the plant in its various aspects as it spoke to her whatever it thought she should know. The Disney version of her story is very good on that point: "Listen to the wind," she sings. And she goes to talk to Grandmother Willow, and follows the Old Tree's advice in making her choices. If you asked her, "Who are your teachers?" she would as likely have answered, "Tree, Wind, Tide, Rain, Corn, Raccoon," as "the Elders" or "the Council of Wise Women." In that sense, the Old Knowledge is indeed an "Oral Tradition."

There is a difference between reading a book and reading the flight of geese, the sough of wind in a certain stand of trees, or the changing rhythms of a stream, however. The one is basically an external act, while the others are inward. It is important to realize, though, that *inward* doesn't mean "inner." It's more that there is a place within one's being—one's center, not one's brain—that connects to the other world, the world of the mystery, the *manito aki*. It is this latter kind of focus that Pocahontas was raised and superbly trained in, and it was her educated skill, along with what must have been an extraordinary talent, that positioned her to play her role as sacred spy and ceremonially empowered diplomat.

The differences in these various ways of accessing and processing information are well described by the physicist David Peat. His explanation neatly sidesteps any need to characterize one view as "correct" and the other as "incorrect" while it helps us understand that there are many "truths," many ways to be accurate. In this view, accuracy depends on what you are describing and the context you are operating within.

In his excursion into Indian Country, Peat draws extensively from his knowledge of quantum physics. In his book *Blackfoot Physics,* he discusses his older colleague David Bohm's idea of the implicate order. He begins by narrating a conversation he had with the traditional Blackfoot James Youngblood (Sa'ke'j) Henderson. Henderson was instructing Peat about a certain gourd, explaining how it "contained the whole world." Henderson begins by noting that Bohm "argued that while the classical physics of Newton described what could be called the surface of reality, by contrast, quantum mechanics has forced us to move to deeper levels of perception of the world." The implicate, or "folded," order that Bohm studied

> suggested that, in its deepest essence, reality, or "that which is," is but a process or movement, which he calls the *holomovement*—the movement of the whole. This flowing movement throws out explicit forms that we recognize through our sense of sight, smell, hearing, taste, and touch. These explicate forms abide for a time and we take them as the direct evidence of a hard and fast reality. However, Bohm argues,

this explicate order accounts for only a very small portion of reality; underlying it is a more extensive implicate, or enfolded, order. The stable forms we see around us are not primary in themselves but only the temporary unfolding of the underlying implicate order. To take rocks, trees, planets, or stars as the primary reality would be like assuming that the vortices in a river exist in their own right and are totally independent of the flowing river itself.

For Bohm, the gourd that Sa'ke'j Henderson carries is the explicate or surface manifestation of an underlying implicate order. Within that implicate order the gourd enfolds, and is enfolded by, the entire universe. Thus, within each object can be found the whole, and in turn, this whole exists within each of its parts.[11]

Neither Pocahontas nor the most learned elders could have described it better.

It should be clear that the Oral Tradition is an outgrowth—an expression, if you will—of the implicate order. Attempts to interpret it as an artifact of the explicate order or as a phenomenon that belongs to it are doomed in more ways than one: they doom traditional peoples and their varied civilizations, along with the entire biosphere (the planet and every phenomenon on, around, and within it), implicate and explicate alike.

The possibility that Pocahontas was an agent of manito intelligence, however it is described, can be explored via abduction narratives and the place they occupy in the Oral Tradition when the exploration remains conscious of Henderson's application of Bohm's theory to indigenous thought. There are a plethora of abduction narratives scattered across the Western Hemisphere. I can offer only a tiny sample here.

Some of the more illuminating narratives in this regard are the Keres stories about a Person named Kochinnenako (Yellow [Corn] Woman). In one of these, which I have seen titled "Sun Steals Yellow Woman," a married woman went to fetch water and the sun abducted her. Her husband, whose name was Stilimo (Dance Shells), grieved so piteously that Old Spider Woman came to his aid. She instructed him not to think about his loss, but instead to go the place where the sun comes up. "There are two roads," she said, and instructed him to take the old road, because the new road was dangerous. Someone on that road would kill him, she warned. He went toward the sunrise house as she instructed, but when he came upon the fork in the road he decided that the road she had said he should take looked difficult, while the other one looked a great deal easier. Ignoring Old Spider Woman's instructions, he set off on the better road and soon came upon Whirlwind Man's house, where the mother of the absent Whirlwind Man prepared to feed the guest. When Whirlwind Man returned and caught sight of Stilimo, he attacked. Whirlwind Man's mother tried to separate them, but in the end Stilimo killed Whirlwind Man. Seeing her son dead, his distraught mother begged Stilimo to restore her son. "Hurry, hurry. Press on his stomach, see if you can revive him," she urged. The bereaved husband obeyed and Whirlwind Man was soon conscious. "Don't fight with this poor one anymore," the old woman scolded her son. "It is because of him that you are now alive." Her words shamed Whirlwind Man, causing him to agree to take Stilimo where he wanted to go.

And so, Whirlwind Man picked up Stilimo and carried him to the high cliffs where Sun's house was located. He told the husband to hurry and get his wife. "Sun is out hunting," the supernatural being explained. Doing as he was instructed, Stilimo found his surprised wife grinding corn. "How did you get here?" she asked. "Nobody comes here because it is too dangerous." They left hurriedly before Sun returned, and soon met up with Whirlwind Man where he was waiting for them.

Shortly after their hasty departure, Sun returned and, finding his captive wife gone, made after them. But they had come to Whirlwind

Man's house safely, even though Sun had shot an arrow at them. His shots fell short because she was pregnant, and, luckily for the fleeing couple, no spirit man can harm a pregnant woman. At Whirlwind Man's urging they rushed to their own home, Sun pursuing and shooting at them all the way. They reached their own home un-harmed, Sun close behind them. Defeated, the magnanimous Sun told Yellow Woman that her child would be a leader of all the people.[12]

Another abduction story, this one from the Lenape, an Algonquin nation that is closely related in dialect to Pocahontas's nation, the Powhatan, concerns a young woman who was taken by a young man of the Thunder people. This gorgeous young woman was in no hurry to marry. Years went by, but still she remained single. Then one time a good-looking stranger showed up, and after a while they struck up a friendship. Finally she went away with him, itself odd as men usually marry into their wife's lodge. Their journey was very long, but finally they came to a vast body of water, one of the Great Lakes. The man went right down under the water, much to the woman's surprise, but, obliged to follow, she submerged. To her astonishment, once she was deep underwater she found she could breathe water, just like air.

After traveling another great distance, they finally reached the young man's house, where his old mother, whose house it was, of course, recognized the girl as mortal. "She can't live here," she protested to her son. "Take her home." But the youth wouldn't be swayed, and they remained together beneath the water as husband and wife.

They lived quite conventionally for some time, but one night the woman awoke suddenly to the sight of a huge snake. Frightened, she fled the house, only to be met by her husband. He wanted to know why she was running away. She said she wasn't, exactly, but had been frightened by a huge snake she saw near where they slept. Thus she assured him.

Her husband reassured her that it was no snake, but only his clothes. But the same thing happened over and over until she resolved to escape. Whenever she was out she would wander farther and far-ther from their place, learning the lay of the land so that when the time came she could make good her escape.

One day she set out soon after her husband had left for the day and went a long way. But then she heard hissing and the sound a snake makes when it moves, and then her husband appeared. She told him she had been exploring the countryside and accompanied him home.

The next time she tried it, she heard the hissing sound sooner. But she decided to call on her friend, her dream helper Weasel, and ask for his help. Immediately Weasel appeared and ran into the snake's mouth. He moved quickly to the snake's heart, cutting it out of his body. At this reprieve, the woman made for the shore, but when she emerged the Thunders were waiting. They carried her up into the air and she realized that she had been underwater all that time. As they carried her, the Thunders kept rubbing her body, and at every rub numerous little snakes dropped from her into the lake, until at last no more snakes dropped and she was restored.

Soon the Thunders carried her back to her village, and she recounted her experiences to those who had gathered to greet her. She told them that she couldn't stay with them, but must now live with the Thunder people. She would come to visit from time to time, she reassured her family, and they would be alerted to her arrival when they heard a cloud rumbling and making a rolling sound. That was the noise her clothing would make, she explained.[13]

Abduction stories can be found throughout the Oral Traditions of many countries. They form a part of every tribal, national, and language group I know of. Sometimes the abductees are men or boys, but more often that not they are women who possess some unique characteristic. In general their characteristic can be summed up as somehow deviant, somehow wayward, or somehow "outside the loop." In some cases this uniqueness might be caused by some sort of tribal dislocation such as war, famine, drought, flood, or other symptoms of cosmic disruption experienced by the community; in others it comes as a consequence of her own nature. She may be curious, spirited, mischievous, as Pocahontas was said to be, and as her name evidently indicated. She might be unwilling to marry. She might have made a wish without realizing the consequences, then discovered to her dismay that she was getting more than she'd bargained for. Frequently she is enslaved by some foreign people, as were Malinalli (taken by the Mexica) and Sacagewea (taken by the Hidatsa). At once, or perhaps later on, the heroine is taken by the supernaturals and thus becomes part of the implicate-order narrative.

Sometimes she is abducted for no "reason" at all, an event that was not all that infrequent in the turmoil that seems to have visited much of indigenous America before the arrival of the Europeans and after (whether you put that first contact at the Viking settlements up north, the Buddhist monks in Mexico, or the Spanish-Portuguese landing of Columbus in the Caribbean). However the abduction comes about, the salient theme of every such narrative is the result in a larger-than-human context. Indeed, the Sky Woman narrative that formed the basis of Algonquin theology (thea-logy), and which includes a variant on the theme of abduction, can be understood only in those terms. In fact, the presence of a female figure embedded in the narrative of the event, whether "historic" or "legendary," can be taken as proof that the event was of supernatural origin and significance.

What this implies is that the supernatural involvement in human affairs is key to the mythic nature of traditional indigenous narrative, American Indian and worldwide. The mythic cycle is not actually a story cycle, nor is it a ritual cycle as present scholarship currently holds. A mythic narrative is a report on some particular person's journey, her experience in Mythic Time, a place as real as the mundane we moderns believe is all there is. Mythic Time is a space; it is a place *between* places we can identify because of their relation to our (explicit) location on the space-time "continuum." *Between* may be somewhere in the quantum continuum that physicists have explored mathematically. It is certainly within the implicate order that Bohm theorized, and that Peat connects with the Blackfoot Tradition. It is, if you will, a quantum fact, and while not of the mechanistic, deterministic, classical explicate order that has framed Old World thought since the orthographic revolution of ancient Greece, its factual nature is testified to by every people, every tradition, and every literature of the premodern world. The ceremonial reenactment of that mythic event is analogous to a scientific experiment that replicates an earlier, original experiment. Through it the event itself can be accessed; its effects can be reproduced in real time to the benefit (or harm) of the community. When the dancers at Santa Ana Feast Day in 1972 danced the pueblo, including the visitors, into the rain, they in effect replicated[14] a Mythic Time event in which the *shiwanna* come to the village and bring rain. When they did so, they (and we with them) went *between* long enough to return with the gift of the *shiwanna*. When the Pawnee reenact Sky Woman's fall, they cause the same conditions to reemerge from the implicate order and make themselves felt in the explicate order of our mundane world.

Sacred abduction narratives are not confined to Indian America. The abduction of someone named Helen can be found all over France and Scotland; I have even run across it in England's West Country. I think that what is being referenced in these narratives is that an ancient sacred ceremony in which a female figure who is liminal—that is, somehow on the threshold between mundane and sacred, explicate and implicate—is abducted by certain powers. This event signals a shift in the order of things: whether this shift is directly caused by human activity or by an intersection between human and supernatural in that time/space between is a matter that traditionals and moderns view differently.

Guinevere of Camelot fame is the heroine in an ancient abduction tale that made its way into the French and English romance tradition. The original abductee's name is lost in the retellings, because Queen Guinevere, like Helen of Troy, was a familiar figure who assumed a mythic identity. In those traditions, the occurrence of the abduction of a figure named Helen or Guinevere in a "historic" narrative sends a signal to the informed listener that "something sacred has occurred" at the time and place the event is said to have taken place. One can argue, as classical scholars do, that such stories are local riffs on the old Helen of Troy theme, but that begs the question. After all, there are thousands of abduction narratives in the ancient Greek, pre–Greco-Roman, and European mythic traditions.

It is pertinent to our quest for the identity of our historic abductee-hero Pocahontas to ask why the tale of the abduction of a central female figure leading to a major transformation in her community is so widespread. Is Pocahontas's abduction and ensuing life among the aliens the major reason she finds herself the heroine of Smith's tale and the darling of the English moneyed class? Smith, after all, was an Englishman, and conversant in English lore. The romance of King Arthur stories—and they were ubiquitous during the Middle Ages—was as vital and alive at the time as it is now. Argall's, Dale's, and his co-acting Governor Sir Thomas Gates's spin on her presence at Jamestown are of particular interest since her abduction may have been more an English public-relations ruse than a fact, about which more will be said later. Nevertheless, the abduction element plainly signals, in Indian Country and outside of it, that a supernatural sequence is afoot.

One thing about these matters—implicate/explicate interaction in traditional societies—is that they are frustratingly paradoxical, perversely eluding analysis and its ever-present companion, deductive logic. It might be said about narratives from the old traditions that what some physicists refer to as the X-factor—the unpredictable event that can randomly occur—is one of their chief characteristics.[15] One could also argue that the Oral Tradition reflects life itself, which is why it defies logic, analysis, and linear modes of discourse.

What is being signified in these narratives is the role played by certain unique female people during "transformation time," a part that is critical to the maintenance of the tradition, which in turn can best be defined as the interface between the implicate and the explicate human orders. It is this role—the shaman-priestess, the medicine woman—that Pocahontas played in the genesis of the United States of America (known to the English of her time as Virginia). Were her story embedded in another narrative from a more "bardic" time, she might have borne the name "Helen" or been assumed under the name-title "Guinevere." As it is, perhaps because the explicate order requires its own mode of narrative to maintain itself, to us her name is Pocahontas. If we thought, remembered, and spoke in implicate-order terms, she would be known as Matoaka, her given name, or, more likely, as Amonute, her name as a sacred high-degree initiate of the *midéwewin,* the Medicine Lodge Sacred Society. Amonute would be the most appropriate name for her in her sacred—that is, world-renewal ceremony—role.

The abduction of Pocahontas by Samuel Argall in 1612, which is quite revealing about the strange, yet all so twenty-first-century con-flation of piety and greed, leaves questions about Pocahontas and her motives in boarding, seemingly oblivious, an English boat that set sail for Jamestown with her aboard.

Whether they be the manito or kachina of Algonquin or Pueblo Na-tive America, the Gentry ("Faeries") or elves of Europe, the ginn ("genies") or angels of the Middle East, or the gods/goddesses of an-cient Greece and Rome, often as not agents of the implicate order, the Great Mystery, are abductees. The heroes of these epics are pre-sented as pawns of the greater powers that seem to rule human affairs, or so narratives from every human community since time immemo-rial assert. They are usually gifted in a variety of ways: attractive, phys-ically fit, strong, even mighty. They are often wiser, wittier, and more profoundly aware than their contemporaries, and for whatever reason (and the reasons vary), they are singled out by the powers of the im-plicate order. Sometimes their story is tragic in all of its details, as in *Oedipus Rex, Hamlet,* or *Camelot.* Sometimes, as in the *Odyssey* or the *Iliad,* the story ends well for the heroic victim of the powers-that-be. Most often, it's a mixture of hilarity and wisdom, tragedy, loss, grief, self-discovery, and love eternal. Sometimes the story is about the deepest spirit of a nation, and such is the story of Pocahontas.

When viewed from a mythic perspective, the question of whom these women—all abducted, all taken hostage, all ordained for great-ness on a global scale—worked for as spies, agents of change, and emissaries can be readily answered. Their mission was defined, ignited, and energized by those forces or powers that lie behind, beyond, and beneath the mundane. They did what they did because they were *how* they were, because that was what time it was, and because their personal characteristics, combined with their training, social conditioning, and the astronomical-quantum standing wave that *was* the time/space they moved in, made it so. It wasn't something for which one can volunteer. It was because it was—because these women were, if you will, born to be agents of change.

Pocahontas, Malinalli, and Sacagewea, historic to be sure, were also mythic because they lived within the mythic as we live within the technological, and because their actions echo down long past their small life spans and beyond their paradigm into ours. The same can be said of Helen—of Troy, Scotland, or England—and of Guinevere and other legendary or mythic figures caught in abduction events. The meaning of the events from the point of view of the Otherworld is not, perhaps, as straightforward as it may seem to us modern folk. To us, the taking, holding, and exploitation of women (or of anyone, male or female), for any reason, is criminal, at best. The problem with such categorical judgments when applied to the Other Realms is that those realms don't necessarily operate in ways that are remotely familiar, or comprehensible, to mortals. This is not a matter of "cultural identity." This is a matter of orders of existence that physicists and shamans alike examine, explore, and strive mightily to document and perhaps understand.

In a charming novel by the Canadian writer Charles deLint, *The Little Country,* a tale about the Oral Tradition's workings in modern Cornwall, one of his characters describes the Native way as clearly as I have seen it described. The speaker, one John Madden, is trying to tell his business partner what it is like to know the old way, a knowledge he seldom lives, for all his knowing, which makes him the villain of the piece. Still, villainous or otherwise, his words bridge the gap between present-day understandings and those of the a-historic peoples:

More than power and glory, [they see] a mystery, a kind of mystical presence that grows more obscure the further you follow it, but each step you take, the more your spirit grows. . . . But best of all . . . you know the mystery will go on, unexplained, and you can keep following it forever. Past life. Past death. Past whatever lies beyond death.[16]

One sees that exact look in the eyes of old people at the pueblo.

Perhaps the elders from the pueblo fixed their gaze *between* when they spoke about the past, entering in that moment the mythic dimension where resides the tradition. It was/is in that time/space that our agents Pocahontas, Malinalli, and Sacagewea worked, and it is that fact that gave their lives and their acts the qualities of the sacred and the mythic. Suffice it to note that whatever goes on in that mysterious realm where women are taken, transformed, and often as not given gifts far beyond those usual to mortals is of an order beyond politics and law as we understand them. The abduction narratives themselves can throw a great deal of light on otherwise obscure matters, the ones about the supernaturals and their goings-on.

Essentially, the story goes, in its barest outlines, something like this. There is a maiden who, as it turns out, is of significant note among her people. Some overwhelming force—call it conqueror, caveman, star husband, holy people, manito, Dracula, E.T., or God, as you will—kidnaps her. She is taken to wherever, where she is made to eat something that will bind her to the force forever, made to take up modes of speech, dress, and worship that will have the same effect, or made to fall madly in love with the abductor, perhaps marry him. Eventually, after a number of twists and turns in the plot, she returns to her people, village, nation, or husband, and brings with her some invisible gift that results in an eternal change in the condition of her people. Then she goes back to the world of the Invisibles—call it her death, the madhouse, enslavement in her husband's palace, or the convent. The change is wrought, and nothing is ever the same. The end.

More often than not she is complicit in some way in the abduction. Maybe she falls in love with Paris. Maybe she wants to get back to the alien nation. Maybe she is bespelled, beguiled, or a silly young thing who thinks glamorous adventures await her. Maybe she marries a mad sorcerer because her dead father tells her to. However she conspires with her fate, there is usually the suggestion that she went willingly, if oblivious to the larger patterns implicit in her act.

So one bright, sunny day—or overcast, cold, and cloudy, who knows?—the now officially mature medicine woman and agent of change Matoaka (Pocahontas) is passing the time with some friends in the land of Patawamack, which lies near her own. Now, these friends are getting on well with some strangers her own people are a bit ticklish about, but she wants to see these strangers. She is rather determined, determination being her most striking characteristic, and she has an adventurous streak a mile wide. So you can bet she loiters about with her friends, hoping to catch sight of the aliens with whom she is enamored.

As it happens, her friends are hosts of one of the bearded aliens. It seems they have an agreement with the visitors: the strangers will play host to some of the Patawamacks and vice versa. Knowing this, and having learned through the moccasin telegraph that the strangers are out of food, again, the beautiful young agent knows it's about time for circumstances to break her way.

And sure enough, there comes the strange boat that looks more like a great waterbird than a mere boat, like in the stories she loves to think about, especially when they have feasts and sacred dances about them. She has had a favorite daydream about how she would fall and fall in an empty forever space until the great winged birds caught her on their magnificent white wings and deposited her in an entirely new world, one that she could shape and make, one that would be the best, most lively, fun-loving, tradition-spurning place she could imagine. In the world she'd make, she used to pretend, there would be dancing and feasting and great fires reeking of sacred tobacco smoke, and elders laughing, and young people flirting and wandering off to be alone in the cornfields, and *hobbomaks* everywhere that would transport a girl just about anywhere, into any time at all.

So, one can just imagine. Here comes the great white bird ship, and her chance to steal away from her duties and her burdens, her young husband and her ceremonial responsibilities, her chance to complete the truly important task her real vision and her spirit teachers directed her to complete. Of course, she goes for it. Wouldn't you? After all, she is sixteen years old, a married woman, and all grown up. She knows what's right and what's wrong: right is enacting her vision; wrong is pushing it aside.

Of course, her abductors don't know they are actually the means—and ends—whereby the requirements of Dream-Vision culture, as they apply to her, are to be made real, part of the explicate order. She knows that they have their own agenda, and she has a plan about that. She will do as they ask; she will undertake to learn whatever they require, speak well, be always in her dignity and power but amiable, compliant. There's a lot more at stake that just her wanton whims, after all.

Never mind they think they're gonna get food, guns, and what-all for their pains. Never mind that they hope to cow her "father," who isn't her father, but there's no point in pointing that out to them. They'd never believe a young thing like her anyway. Besides, she's a "salvage," as Smith writes, and she has "salvage" agendas of the manito and therefore of her own. She is savvy enough to know how ugly they would get if they knew she, with the knowledge and concurrence of the Council of Elders, was having them on.

Thus, back at historic *tsenacommacah,* where what seem to be folktales to those of us on this side of the implicate/explicate divide were no more than everyday events, a few years after she was chosen as spirit-guide sponsor of the aliens' Virginia Company, Pocahontas was out with her group—her "attendants," as the English records have it. There, accompanied only by the local leader Japazaws ("King of Patowmack," as Smith identified him, but probably a minor chieftain) and his wife, she entered the next phase of her ceremonial program journey. She went to the fatal enemy on an English ship under the command of Samuel Argall. The record makes it clear: she went aboard at his invitation. However duplicitous he thought he was being, however he believed that her companions were in collusion with him, contrary testimony suggests that that all was not as Argall supposed. It was a sting operation, and the Powhatans were conning the English, who thought that they, being superior to the "salvages," were in control.

According to records of the time, Argall sailed to Pastanancie, where he heard that Pocahontas "lay concealed . . . hard by Patowomack, thinking herself safe, and unknown to all but trusty Friends," as Argall reported. He had gone to Pastanancie to pick up a young Englishman who was in the keeping of Japazaws. Called a "hostage," the term as the English used it at the time meant an agreed-upon exchange of young men rather than a captive held for ransom. This young man, Ensign Smith, evidently told Argall about the hiding princess. Why she was in hiding, or from whom, is not made clear; indeed, that Argall took the story literally suggests misunderstanding at best, chicanery in a worse case. Company secretary Rafe Hamor recorded the testimony of the ensign, who implicated Japazaws and his wife, accusing them of "venality." Argall himself believed they sold Pocahontas for a few pieces of silver or similar trinkets—another familiar story in the English tradition.

There was another mystery here, in addition to why and from whom Pocahontas might have been hiding; she was married, after all, to a young Indian named Kuocum, and served as a trader for the Great Council, acquiring corn and other commodities for the stores at Werowocomoco and perhaps Orapaks. So unless she had run off with the goods, she couldn't have been hiding from Wahunsenacawh. Hiding from such a formidable *mamanantowick* would have been foolish as well as futile, even for one as versed in the magical arts as Pocahontas. That leaves the notion that she was hiding from the English, but that makes no sense, either. She was their friend and ally, or had been until the latest spate of open conflict.

Since taking on the responsibility of Nantaquod (Smith) and his tribe (the Virginia Company James Fort contingent), she had often visited Jamestown. She headed delegations that brought gifts of food while serving as an emissary and informant for the Great Council. As the English understood it, she had even sided with the aliens against her "father" when she warned Smith of an "impending attack," in the winter of 1609, as Smith's account and other records show. That her intervention might have been deliberately designed and implemented by the Great Council seems to have occurred neither to John Smith nor to subsequent historians.

In his report to a friend back home, Smith had declared his profound faith in the young woman. "Wee now remaining in being in god health, all our men wel contented, free from mutinees," he wrote, adding "and as we hope in a continual peace with the Indians." That he based this optimism on Pocahontas seems clear enough. He trusted her sufficiently to make concessions on the strength of her visit to the colony to see to it that some Indian prisoners were freed. She had gone as ambassador, Wahunsenacawh having assured Smith that she represented the council in that regard, and would remain above the negotiations conducted by the men's families and their captors, ignoring the hostages but overseeing their freeing. She remained in the English fort, "not taking notice at all of the Indians that had beene prisoners three dais till that morning that she saw their fathers and friends come quietly, and in good termes to entreate their libertie." It was into her custody that the men were given. Smith "delivered them into the hands of Pocahuntas the Kings Daughter . . . for whose sake onely he fayned to have saved their lives, and gave them libertie" to demonstrate their appreciation to the Powhatans for sending her as emissary. They also gave her some "trifles as contented her," which Mossiker lists as "Venetian beads; mirrors . . . hawk's bells [used by falconer's and fastened to birds' feet]."

Smith, in familiarizing himself with the Powhatan language, had an exercise book in which he wrote down sentences, in English and phonetically based Powhatan, which he shared with the girl. Thus she and he learned one another's language, a task that would help Pocahontas mightily in the later stages of her enterprise.

"Bid Pokahuntas bring hither two little Baskets" went one of his entries, followed by the Powhatan: "*Kekaten Pokahontas patiaquagh niuagh tanks manotyes.*" He made entries during hostage trading, an event that served both Smith and Pocahontas well enough. Had Smith's ambition to remain in Virginia, on his own if necessary, been realized, the language lessons would have served him even better.[17] Smith was forced back to England in 1609, a turn that must have taken him by surprise but should come as no shock to those who read manito workings into otherwise seemingly accidental but oddly timed events. Considering the duplicitous actions of John Smith, dealt evenly to the English who impeded him and to the Powhatans, despite the latter's many attempts to deal honorably with him, or the former's attempt to put a legal halt to his machinations, the outcome was all but guaranteed.

In the case of the hapless Mr. Smith, a man given every opportunity to distinguish himself as the first successful English envoy to the Indians of North America by those same Indians, the pattern emerges clearly enough, particularly when one notes how heavily influenced were those opportunities by the manito energy. The depth and profundity of his missed opportunity will be made explicit years after his removal back to England when he again meets his "ideal Indian Princess" during her stay there. He will once again miss the point; whether perversely or out of genuine ignorance remains to be clarified.

Briefly, John Smith, an enterprising individualist on every score, had jockeyed himself into a position of power in the embryonic settlement. By dint of a judicious mixture of duplicity, mendacity, and tenacity, he led the Powhatans to understand that his group had no intention of remaining in the area long. (They had gone adrift, he assured them, and would leave soon.) Simultaneously he was confronting the English, to the exasperation of the company bigwigs as well as the detriment of his fellows. He began his New World experience in the brig, where he was being held pending trial on charges of mutiny. Freed, he bumbled into a people's effort to stave off their prophesied destruction, and perhaps it seemed for a time that their last-ditch effort at survival might pay off. Should the effort to Powhatanize the boat people fail, the manito had a backup plan.

But until it was certain that Smith couldn't or wouldn't serve in the office to which the manito and the council appointed him, his luck was almost as incredible as his ambitious misadventures. At one point he was once again up on charges for the murder of two men, Emry and Robinson. The men had died when Smith was with the alliance. Many of the men sided with Smith, who had built a strong following among those who like himself were of common origin. But the council executives—John Ratcliffe, the new president of the council; Gabriel Archer, who, having studied law at the prestigious Gray's Inn, served as prosecutor; and Captain John Martin, who belonged to the upper-class faction—wanted Smith's head. They very nearly succeeded. Just in time the expected ship arrived from England bearing Captain Newport, who was set to take control of the company—and crown Wahunsenacawh on King James's order (which had taken some fine maneuvering to gain) while he was at it. Smith's head was in the noose when the ship hove into view, and he got off. Nothing daunted, Smith continued his own machinations, heartily disapproving of Newport's "liberal" ways with the Indians, with whom Newport bartered much more generously than Smith thought necessary. He seemed determined to shape the enterprise to his liking, and for some time the manito seemed in support of his dream. But he was English, and strong-willed to boot. He had no patience with social niceties, whether Powhatan or English, and he dealt with both groups as he chose. Usually he chose duplicity: double-dealing with the best con men of that or any era, he was a seventeenth-century prototype of Dickens's Artful Dodger. He gave the *tsenacommacah* the assurance that English presence was temporary, a deceit that increasingly lost credence with the coming of each English ship to Jamestown and increasingly led to the Powhatan Alliance's hostility toward the aliens, whom they grew to view as what they were from the beginning: invaders.

Smith was aware that English settlements were not on the Powhatans' wish list, a matter that Wahunsenacawh, Opechancanough, and others had made clear. That being so, he went out of his way to assure them that he, and the people they presumed him "father" of, were not to be in the region long. Of course he was well aware that the Virginia Company, enterprise of the crown and the City of London, had plans that were quite the opposite of what the Indians desired. No fool he; fearing that the small contingent would suffer the fate of the lost first company at Roanoke, Virginia, several years before, Smith didn't reveal the bigger plan to his erstwhile hosts.

But although the English, ill suited to the rigors of life in the wilds, went through some difficult times, they hung on. The *Sea Venture*, which brought John Rolfe as well as Smith's reprieve, had broken apart in a storm at sea near Bermuda. Those who had survived the shipwreck rebuilt the *Sea Venture* into three smaller ships, finally arriving at Jamestown in time to head off Smith's execution. Meanwhile, as time and energy permitted, the English kept improving their fort, buildings and fortifications alike. Much of this effort occurred under the strong direction of Smith, who became head of the company when the company's first council president, Captain John Ratcliff, and Gabriel Archer, Smith's personal nemesis, departed for England in 1608, leaving the field clear for Smith's ascendancy to the council presidency as he'd hoped from the time the company had left England. True, his English adversaries had gone armed with charges of dereliction of duty, accusing him of mismanaging the colony and putting the company at risk. They had earlier charged him with the murder of two of the men, for which he had been tried soon after his return from his original *tsenacommacah* adventure.

Nevertheless, with Ratcliff and Archer dispensed with, Smith's way was clear. He had big plans for himself, did John, and the manito seemed to be on his side. Had he taken advantage of his good fortune, his fate might have been quite otherwise. Rather than sinking into frustrated ignominy in England, unable to return to Virginia or get involved in another adventure elsewhere, he might have been the friend of the Indians he is reputed to have been, rather than the deceitful operator his testimony and, more important, the charges Pocahontas leveled against him show him to have been.

The fatal blow to his dreams landed in July 1609, when Captain Samuel Argall arrived with word that the company was sending more colonists and soldiers out. Smith received the announcement from Argall, learning that a second charter had been issued in May of that year, which should have been welcome news. However, the organization of the company had been revised: the colony would be separated from the North Virginia Company, and the council would be abolished. This meant Smith would soon be out of a job; the new governor would be Thomas West, Lord De la Warr (after whom the state would someday be named), and the office would be designated a lifetime position. The new company, the Treasurer and Company of Adventurers and Planters of the City of London for the First Company in Virginia, would govern Jamestown via a supreme council based permanently in London. West would be the vice president of the Jamestown office, so to speak. The company was optimistically sending nine ships of colonists over; they would arrive soon after Argall.

Receiving the bad news, Smith determined to set up his own holdings away from the English settlement. He decided to settle on land near some falls in Powhatan territory, and would name his place "Nonesuch." It was an ironic choice, for none such ever occurred. Smith, heading back to retrieve some materials to build his place, was caught in an accident. A bag of gunpowder exploded, seriously injuring him.[18] So severe were his wounds that his friends put him aboard a ship returning to England, putting an end to his adventures. In this way he went from adventurer to adventure writer, long heralded as the "Father of American Literature."

American literature—at least in its popular dimensions in print, cinema, and television, as well as role-playing and Internet games—is noticeably marked with Smith's cast of consciousness, as is American culture in many of its public dimensions. Some suggest that for many chroniclers of the Smith legacy, the tragedy is that he didn't marry Pocahontas, who, as they believe, loved him deeply. Had he done so, their laments seem to say, he could lay biological claim to the title "Father of America" as well. But in truth, the best that can be said of the man is that his ability at self-promotion, coupled with his taste for

adventure and hyperbole and his active pen, made him the major figure in the Jamestown adventure; he may yet be known as the "father
of Madison Avenue."

Leaving the matter of her "hiding" unanswered, it is more likely that
on this summer day Pocahontas had no particular reason to hold back;
indeed, it is very likely that it was just this occasion she was waiting
for. When invited to accompany Argall and his crew to Jamestown,
she boarded his ship, with no expectation other than a pleasant boat
ride downstream and a continuance of her duties as spy and perhaps
emissary. Because of worsening relations between the two communities, after the sudden departure of John Smith her contact with the
English had been severely curtailed. She needed to get back in: the
council needed eyes and ears within the alien enclave.

Of course, the English had a hidden agenda of their own. As they
saw it, the "Princess," beloved daughter of the "greate king Powhatan,"
was delivered to Jamestown a hostage. At first, or so their records
state, their plan was to trade her for some Englishmen the Powhatans
held. The company was certain their ploy would be successful, for
wasn't Pocahontas the much-doted-upon daughter of the fearsome
Powhatan? But also at work was the Puritan clergy's interest in converting Indians to their cause, saving heathen souls and coincidentally
expanding their influence over the direction the English church
would take.

Taking a modern view of the Pocahontas story, Charles Larsen
gives us the point of view that many radicalized Native people sub-
scribe to: that she was a traitor to her people, favoring the English
over them. Such a view, of course, depends on a post–Civil Rights
Movement analysis that is itself based entirely on the English spin;
hers is one of the classic examples of the adage "to the victor belongs
the spoils"—especially the history. As Pocahontas is herself voiceless
in this concoction of historic distortions, political ploys, and propa-
ganda of earlier times, we do not know what her version of the events

might have been, as Larsen pointedly remarks. Somebody has the record, though: the ancestors, the manito, Pocahontas-Amonute herself all do. It's a matter of seeing the world from their point of view and reading the evidence left us in that light.

Larson prefaces this part of his lively discussion of the Pocahontas of drama, melodrama, fiction, and biography by observing, "At its very origin, the Pocahontas myth is rooted in insidious racism."[19] He notes, regarding her "betrayals"—included her ongoing aid to the company, her Christianization and marriage to an Englishman, her voyage to England, and her wish, or so it is said, to remain there— that, taken hostage, she may have been forced into conversion and marriage, since, as is generally held, the great king Wahunsenacawh didn't see fit to meet the Virginia Company's ransom demands that would have freed her.

In the English world, a king would pay dearly to ransom his favorite, the "apple of his eye" and his "chiefest darling," but no such luck for Pocahontas. Wahunsenacawh made some efforts to parley with the English aliens, but when they wouldn't trade with him fairly as a tribe and thus a member of the *tsenacommacah* he led and to whom these same aliens owed allegiance on Powhatan terms, he gave up. When Wahunsenacawh and his allies retreated, they left Pocahontas either stranded or in place as a "mole," a secret agent who could pass information to the council and the Midewéwin while acting as the human "eyes" and mind of the manito. The interpretation of her placement among the English favored by historians and novelists rests on acculturated beliefs taken for fact. Earlier interpretations of the facts of her life are similar to current ones, only they didn't explain her actions as the result the "Stockholm syndrome," but rather describe her as a docile and willing convert to Christianity and civilization, a model for us all, particular American Native people.[20]

But if Pocahontas was "Mischief" personified, involved in not only espionage but a kind of undercover diplomacy as well, if her vision and her job as medicine woman required that she remain among the English and adopt their customs, the *matchacómoco* would not have paid her ransom but left her to function as a "mole" or "sleeper." Because she was agent for the people of the Dream-Vision, *powa manitowinini,* as well as for the manito, over the ensuing months and years most of the Powhatans involved, including Wahunsenacawh, would act their parts, which they did, brilliantly.

THE ABDUCTION OF POCAHONTAS
Engraving by Theodore De Bry
(courtesy of the Virginia Historical Society)

3

Manito Aki / Faerie

Uneven balance
of mountain and dream
We still live in another world
perhaps the interval
                    —Phillipe Jaccottet

True Thomas lay o'er yon grassy bank
And he beheld a lady gay,
A lady that was both brisk and bold
Come riding o'er the fernie brae
. . . . He has gotten a coat of the green greencloth,
Likewise shoes of the velvet sheen,
And till seven years were past and gone,
True Thomas ne'er on earth was seen.
                    —Thomas the Rhymer,
                    Scottish Traditional

Although the arrival of the Europeans and most of the ensuing consequences over the next centuries had been recorded in the Oral Tradition as much as a century or more before they occurred, both the Mexica (Aztecs) and the *tsenacommacah* (Powhatan Alliance) had known for a long while what time was upon them. They had this knowledge not because of copious data gathered from afar, but through the tradition, their astronomers and prophets, and their communication with the manito, all of which told them what would come. No matter how gently or ungently they went into that good night of their time, however much or little they raged against the dying of its light, they knew, and for the most part accepted, that the expected period of great change was upon them as the record, stunningly, holds. Their ongoing relationship to the supernaturals and the "otherwhere" gave them the abilities to relate to what the tradition offered in a living, present way. The process involved is comparable to the time of birth, when a woman, whatever her feelings about her fate, accepts that labor will do with her what it will, and that it might kill her. For most women, what it will be is bloody, painful, and requires some years for recovery.

However the tale is constructed, even when monolithizing the Powhatan *tsenacommacah* into the generic "Native Americans," I think it was very differently experienced by Pocahontas. She—whether in her identity as "Pocahuntas" the "young wanton," as the surviving "Ancients" (the first to enter the *tsenacommacah*) recalled, larking about, asking, "Do you love me not?" on every possible occasion; in her identity as Matoaka, as she was known to her clan as a child and then as a woman; or in her identity as Amonute, medicine woman— was following her Dream-Vision. She was much more than a simple Indian maiden: she was an initiate and a powerful practitioner of the Dream-Vision People, a shaman-priestess in modern terms.

In each of these interconnected but distinct roles, she was remaining true to the path, the Great Medicine Way. Her education, her experience, and the Great Medicine Dance tradition (all three of which included close contact with her personal manito, or *powagan,* as well as other manitos) guided her on her foray into the worlds that she knew lay beyond the *tsenacommacah,* the English James Fort and, later, England itself being only two of them. In the various lands she lived in, she well knew that all manner of events, familiar or shockingly strange, might occur. It would be up to her to follow the path and return to the *tsenacommacah* bearing some intelligence that would provide them a means of transport—mental, social, physical, or otherwise—from their old system into the one that came upon them. I doubt it made much difference to those Old Ones whether she came back in her Pocahontas body or not. As Sealth (Chief Seattle), Squamish/Duwamish speaker, pronounced, "Did I say death? There is no death, only a change of worlds."

In the telling of this Indian woman Pocahontas's life, it is tempting to do so in Euro-American style, whether by valorizing her for being a model of savage Christianhood or scapegoating her as a proponent of conquest and therefore traitor to her people. In modernist discourse historians and anthropologists view her life within a context of genocide and imperialism, of historical imperatives and market forces, of carpenters and kings. Such constructs are familiar to modern minds. Harder, but alone pertinent to the life the Powhatan-Algonquin woman with three, then four identities as Pocahontas, Matoaka, Amonute, and Rebecca, to the biography of the *midéwewin* adept who would become an international figure, is that she herself was situated in a mind-set that made it relatively easy for her to cross from a world we moderns cannot recognize to one we would find more comfortable and known. It is well to keep in mind, though, that the world of Pocahontas was very close in concept and in reality to the world of her seventeenth-century European counterparts.

There were a number of stories in European folk as well as literary and popular fiction well known among English people at all levels of English society. One of them, about Sir Gawain and the Green Knight, was written down by the same person who also wrote down a long Dream-Vision poem titled *The Pearl.* Both works were recorded sometime in the mid–fourteenth century but remained current among educated men and women. Laborers and other unlettered people were also familiar with Arthurian tales. Stories and ballads had

long been sung or told wherever people gathered. Because of the live-
liness of the oral and written forms of these ancient stories, those who
came here at the time were at least familiar with the ideas that were
fundamental to life in the *tsenacommacah,* had they realized it. Unfortunately, stories about fairies, giants, ogres, little people, and elves did
not form a portion of business writing, which is the kind of writing
modern scholars depend on to learn about life and thought in the
early colonial period.

Nevertheless, two streams of thought emanating from England's
pagan tradition flowed strongly among them. These strains were so
clear in that era that they underlay and informed the works of
William Shakespeare, Christopher Marlowe, Ben Jonson, and other
poets and dramatists of the period. This body of thought enabled the
English and the Indians to converse because, despite centuries of
Christian teachings, the two communities shared a basic worldview,
even as their different agendas drove them into lethal conflict.

Not entirely coincidentally, the works of Shakespeare and Ben
Jonson played a cameo role in the life of Pocahontas, for during her
stay in England she attended productions of both. One supposes that
these productions seemed no more alien to her than the masque that
she and her friends had produced and performed for the English had
seemed to them. The main point of interest here, though, is that
Pocahontas's life was as much about life within the *manito aki* or fairy
realm as it was about politics or economics, though these led to conquest by aliens from somewhere beyond the edge of the world, at least
to the modern mind.

In the English fairyland Oral Tradition, the figure of the Green
Man figures large; it goes back into "time immemorial," thousands of
years. This figure, known to most if not all of the Celtic-Germanic
peoples, had various names: he might be Pan, the goat-footed god;
Loki, the Norse trickster; or the wonderful Puck of Shakespeare's *A
Midsummer Night's Dream;* or even the Devil, the great Satan in Christian lore and teachings. His connections are always with the land, out
of which he emerges on occasion, and back into which he fades
when his task is done. In a number of ways, this English story is re-

markably Algonquin, in spirit if not in detail. In many ways it operates as a spirit-world script for the life and times of Pocahontas, though in her case Sir Gawain is replaced by the medicine woman Amonute, whom we recall as Pocahontas. What is not clear in her story or in the tale of Sir Gawain and the Green Knight is the identity of the Green Man, other than that her *powagan,* or spirit guide, and his Green Knight may have played parallel roles.

Moderns inquire, "Who was he, she, or it, exactly?" And exactly, we can't say. But stories of the Trickster-Creator god that abound in human narrative cycles worldwide hint at the role, the function of this being. Attempts to analyze and categorize this figure, whether as Loki, Coyote, or Brer Rabbit, are far less successful in enlightening us than the multitude of stories about such figures can be.

Suffice it to say that the Oral Tradition has an odd capacity: it foretells the future of given individuals with uncanny accuracy. The phenomenon is widespread and well documented in the annals of folklore, and while discounted in modernist chatter, the inevitability of its workings remains undauntedly so. Usually told in terms of established characters such as Yellow Woman, Coyote, Kokopelli, Glous'cap, First Woman, Selu, the Green Man, Guinevere, King Arthur, Sir Gawain, and the like, these narratives chronicle the times or the dream lines familiar to Dream-Vision cultures. Pocahontas was raised and educated in one such group, which took its name from the event that was the most important to them: *powwaw,* Dream-Vision. The story as it appears below is in a form it would have taken for the literate among Pocahontas's English contemporaries. But as it had come from the English Oral Tradition, it would have had a folktale version that would have been known to the laborers. This excerpt is a reminder that the Oral Tradition was alive and still thriving in Europe at the time the Europeans first encountered Algonquins and other Native people here.[1]

The tale of Sir Gawain and the Green Knight is also useful in that it tells an old story that can be found across human cultures and across time. That story is about how the mortal realm we usually inhabit is closely connected, at least at certain times, to other worlds we don't usually encounter.

Gawain, one of the main heroes of the Arthurian tradition, gets himself into a situation where he has to meet with the supernatural giant known as the Green Knight. This being may be one of the forms of the Green Man (Herne, Puck, Cunennos), a mythic being who graces mythic traditions across Europe, and who is said to appear now and again on Chalice Hill and on Glastonbury Tor in Somerset shire, England. He mates with Sovereignty, Earth goddess, in the fabled mythic land called "Logres," both at that time and at present. *Logres* is a way of naming the implicate order in which humankind, critter-kind, plant-kind, and supernaturals of all kinds exist. While invisible to an explicate-order frame of mind, it informs and pervades the explicate much as yeast pervades bread. *Logres* can be thought of as the English form of *manito aki*. Even now, many consider it the real world, although many others would think of it as fantasy island.

According to myth, the mating of Sovereignty and Cunnenos, the Green Man, causes spring, new life, prosperity, and all the good things that keep mortals and their spirit kin in tune. The story of Sir Gawain and the Green Knight is actually part of a vast tradition that likely would, and does, fill libraries. This tale, though, is about the darker side of spring, when the spirit of life is the giver of death, when the hero-king faces his mortality and that of all the earth, to find that out of honor comes renewal.

So old is this tale that it can be traced back to the lore of ancient Egypt, the love between Isis and Osiris, the conflict that arises between the brothers Osiris and Set, and the mending of the ensuing devastation. Isis, the shaman-priestess who is eventually seen as "Goddess," with the help of her divine sister, Ishtar (also known as Astarte), rescues her murdered consort-brother, reconstitutes his body (except for his penis), and magically, with what must be the earth's first dildo, impregnates herself. The water of the sacred river, the Nile, is the source of the seed that quickens within her womb and is finally born as Thoth, scribe of the gods, who must have been the *Pearl* poet of his day. While this ancient Indo-Germanic myth may seem far from Pocahontas and her people, one must remember that it bears an eerie resemblance to the story of First Mother."[2]

While only portions of this greater myth are contained in the story about Sir Gawain and the Green Man, the idea remains firmly in place. The Green Man, the King of Earth's consort, almost enchants the Knight into perfidy, but Gawain's acceptance of his behavior is sufficient to demonstrate his devotion to the code of honor. Contrast this with the actions of John Smith, both during his stay in the *tsenacommacah, Logres* West, and after. Unlike the hero Gawain, Smith is incapable of accepting his role and admitting it, and because of his failure Pocahontas turns away. Just so, the Green Knight, the magical life force itself, would have turned away from Gawain had he not lived by the code of honor that held sway among the Knights of the Round Table. The significance of the English mythic tradition to the story of Pocahontas is plain: she became an Englishwoman by choice. Being who and how she was, she thus became intimately involved with *Logres;* so intimate was her connection to that green and noble land, England, that her body carried her spirit into its Underground—another, more common name for *Logres.*

The tale of Sir Gawain and the Green Knight, like the myths of Sky Woman, First Mother, and others from the Algonquin Oral Tradition, is a narrative version of a shamanic (i.e., magical) ceremony. Adepts, wizards, witches, shamans, priestesses, magi—any practitioners of the arcane who are highly trained can recognize how the narrative is put into practice as a ceremony. The world and the facts to which such stories refer are "sacred"—that is, they are very real, energizing, and influencing events within the implicate and, as a consequence, upon the explicate order of things.

According to the story of Sir Gawain and the Green Knight, it was Yuletide, New Year's night, at the court of King Arthur. The knights and their ladies were drinking and feasting before the Yule log, when, along toward midnight, in came a huge being, one with fierce eyes that had an ethereal luminescence. He carried a correspondingly huge ax, and, crossing the hall, he came to the high table, where he was offered some ale and their hospitality, as was the custom. Now, the listener or reader of the tale knew who he was, for such events at the time were common knowledge. The Green Man was known to frequent such holy occasions, and it was the court of King Arthur's bad fortune to be graced with his mighty presence on that night.

All dressed in green and mounted on a green horse, he rode straight to the high table, and addressed the king. He inquired who was the head of the people gathered there, for he wanted to speak to the hero—which I take to mean leader. His demand was greeted with silence as the assembled stared at this apparition in amazement:

> Then was there great gazing to behold that chief, for each man marvelled what it might mean that a knight and his steed should have even such a hue as the green grass; and that seemed even greener than green enamel on bright gold. All looked on him as he stood, and drew near unto him wondering greatly what he might be; for many marvels had they seen, but none such as this, and phantasm and faërie did the folk deem it. Therefore were the gallant knights slow to answer, and gazed astounded, and sat stone still in a deep silence through that goodly hall, as if a slumber were fallen upon them.[3]

Addressing the court, the terrifying stranger asked who was lord of the place, as he had a challenge to offer. When King Arthur identified himself the Green Man announced that he had come to offer a challenge to the king. He instructed the king to smite him with his broadsword. After that he, the Green Knight, would respond in kind.

Realizing at once that the contest was at best unbalanced and that the king must be protected at all costs, Sir Gawain leaped to his feet and begged the king to allow him, Sir Gawain, to fight in his place. Arthur gave way, the men met, and Gawain struck a mighty blow, severing the visitor's head.

Calmly, the giant retrieved it, and, holding it by the bloody hair, blood streaming from it and his unheaded neck, he addressed the young knight.

Look, Gawain, that thou are ready to go as thou hast promised, and seek . . . till thou find me, even as thou hast sworn in this hall in the hearing of these knights. Come thou, I charge thee, to the Green Chapel, such a stroke as thou hast dealt thou hast deserved, and it shall be promptly paid thee on New Year's morn.

Alas for Sir Gawain. In attempting to save his king from a ghastly fate as the knightly code requires, he brings that fate down on his own stalwart head. Perhaps Gawain's choice sheds a bit of light on the questions raised by Pocahontas's acceptance of Christianity, her marriage to the enemy, and her evident delight in England. In most accounts, her actions are cast in a way that leads one to suspect that her dreadful fate occurred because she had had her head turned for love of Englishmen and English trinkets.

However, there is another way to view her choice: perhaps she was honor bound, as a medicine woman, to go to the English and live as one of them. If so, the most dreadful thing she could have done, from the point of view of any Native society I know of, would have been to go back on her vision, which was more than a vow. For her people then, and for many Indians now, dishonor is not an option. Nor were the English of her time unacquainted with profound commitment to honor, although the "new man" of the time seemed to count honor a kind of commodity, for sale at the right price.

Like the story of Pocahontas, that of Sir Gawain and the Green Knight is a Dream-Vision epic, stemming from a time in England when Dream-Visions were common enough to warrant long narratives recounting them. In the accounts of her from the time and as the centuries pass, one can see in the growing legend of the mysterious Indian princess Pocahontas the emerging outlines of a myth, one that seems to originate far from Camelot and the fabled court of King Arthur and his famed Queen Guinevere, but Pocahontas has herself become part of the Oral Tradition of that England that still recalls Arthur, his lady, and his knights, chief among them Sir Gawain.

One of the events that make the story of Gawain and the Green Knight of interest in considering Pocahontas's tradition and the beliefs it cherished, other than that it is a powerful example of the Dream-Vision tradition of Pocahontas's adopted people, comes when the knight of the Round Table crosses from the familiar world into the world of Faerie—identified as "Logres" in the text. While Gawain's journey in the realms beyond this mundane one may have differed in its particulars from similar journeys that Pocahontas must have made as she became versed in the ways of the *midéwewin* and walked among the inhabitants of the *manito aki,* these same details give us some idea of what she faced in her own training. The continuing presence of *Logres* in England must have provided more than sufficient reason for her, a medicine woman of proven, unfailing curiosity, to remain there a while longer. What medicine person would pass up such an opportunity, especially at such a powerful time as the time of a change of worlds. The poem continues:

> So rode Sir Gawain through the realm of Logres, on an errand that he held for no jest. Many a cliff did he climb in that unknown land, where afar from his friends he rode as a stranger. Never did he come to a stream or a ford but he found a foe before him, and that one so marvellous, so foul and fell, that it behoved him to fight. So many wonders did that knight behold, that it were too long to tell the tenth part of them. Sometimes he fought with dragons and wolves; sometimes with wild men that dwelt in the rocks; another while with bulls, and bears, and wild boars, or with giants of the high moorland that drew near to him.

Perhaps the poet's written record of Sir Gawain's lonely journey foreshadowed a similar one a young Powhatan woman was honor bound to undertake centuries later. For both the knight and the medicine woman, oath and honor bound, what was begun in a sacred season had to be seen to its end. As Sir Gawain prayed for aid in his dangerous mission, Pocahontas doubtless asked for the help and protection of the four directions and a being Roger Williams identified as Squáuanit.

THE WOMAN BRANCHES OF *MANITT-MANITO*

One of the earliest records concerning the spiritual beliefs and practices of the Algonquin were kept by an Englishman and pertained to the Narragansett, a northeastern Algonquin tribe. Some of the words his informants gave that reporter are close to those recorded by Smith, William Strachey, and Robert Beverly. Roger Williams's work with northeastern Algonquins provides some information about Pocahontas's spiritual knowledge.

In *A Key into the Language of America,* Williams, who founded Providence, Rhode Island, tells us that the Narragansett had a profound belief in a pervasive great being they spoke of as manito. Following an entry for "Manìt-manittó wock," which he translates as "God, Gods," he observes: "They will generally confesse that God made all; but then in speciall, although they deny not that English-mans God made English Men, and the heavens and Earth there! Yet their Gods made them and the Heaven, and Earth where they dwell." This startling observation provides a major clue to Pocahontas's mission: in taking on an Englishwoman's identity, which was accomplished when she took instruction and a new name and underwent the Christian water ceremony. It was sealed with her acceptance of an Englishman as her mate. She took these steps in order to gain the *manit powa* of the English and transfer it to her Powhatan medicine women and men.

Williams continues with a few more examples of the use of the word to provide context and various grammatical forms of the term, then adds:

> But herein is their Misery.
> First, they branch their God-head into many Gods.
> Secondly, attribute it [*sic*] to Creatures.
> First, many Gods: they have given me the Names of thirty seven which I have, all which in their solomne Worships they invocate: as. . . .

He proceeds to list several, including the gods of the four directions. He is informed that *Kautantouwit,* the great South-West God, is where all the souls go, and from whom come their staple vegetables, corn and beans—and squash too, I imagine, though he doesn't mention it. Then he continues:

> Even as the Papists have their He and Shee Saint Protectors as *St. George, St. Patrick, St. Denis,* Virgin *Mary,* &c." he names Squáuanit, whom he styles *"The Womans God,"* and "Mucquachuckquand, *The Childrens God."*[4]

In a telling passage, Williams clarifies the complexity of Algonquin thought, a complexity that vexes him. He thinks it's a muddle worse than that of the hated Papists, and his perplexity has long been shared by students of Native linguistics who have followed him.

> Besides there is a generall Custome amongst them, at the apprehension of and Excellency in Men, Women, Birds, Beasts, Fish, &c. to cry out *Manittóo,* that is, it is a God, as thus if they see one man excell others in Wisdome, Valour, strength, Activity &c. they cry out *Manittóo* A God: and therefore when they talke amongst themselves of the *English* ships, and great buildings, of the plowing of their Fields, and especially of Bookes and Letters, they will end thus: *Manittôwock.* They are Gods: Cummanittôo, you are a God, &c. . . .[5]

In regard to which supernatural powers might have accompanied Pocahontas—Amonute in her identity as medicine woman—or which she might have invoked, Squáuanit, the "Womans God," as Williams identifies this deity, would have been prominent. This being, given the name with which Williams identifies her, is currently known among eastern Algonquins as "Granny Squannit," who in many ways resembles Queen Mab, of the English tradition. Like Granny Squannit, Mab, the Faerie Queen, is usually associated with *makiawisug,* "the Little People," who might be identified as "pixies," "elves," "fairies," or even "leprechauns."

In some ways the character of Grandmother Willow in the Disney Corporation's *Pocahontas* echoes her, perhaps deliberately. Granny Squannit is a supernatural female being who helps humans in a number of ways. She is connected to the Little People, and is as profoundly the Spirit of the Land as is the English being known as Sovereignty. The health and well-being of Granny Squannit is the health and well-being of the earth, and its creatures—human and otherwise—are dependent on her for their own health and well-being.

In some circumstances, Granny Squannit resembles the Fisher King, of the Arthurian cycle. The Fisher King has antecedents in Egypt, where the death and remaking of Osiris is a central ceremonial narrative. In all these traditions—Algonquin, English, and ancient Egyptian—the failing health of the supernatural being is mirrored in an ailing earth. In a story about Granny Squannit and the *makiawisug* recorded by the Mohegan writer Melissa Tantaquidgeon and the Abenaki writer Joseph Bruchac, a medicine woman named Nonner Martha is called on to heal Granny—as, in the Egyptian cycle, Isis seeks and finds Osiris and restores him to wholeness. It is this figure, whether it be identified as Grandmother Squáuanit, the Fisher King, or Osiris, whose health and well-being foreshadow that of the earth, the land, and its creatures. In Sir Gawain and the Green Knight it is Sir Gawain who must restore the earth, Herne, the Green Man; in another Arthurian narrative the same Sir Gawain seeks the Holy Grail in an attempt to heal the ailing Fisher King. Some scholars suggest that the sacred figure of the Fisher King is somehow connected to the Christ figure in his aspect as the crucified God who is ever dying. One might read the Green Knight, holding his bleeding head by the hair, as the earth in its dying phase at midwinter. In a narrative of the Mohegan tribe (northern Algonquin), it is Granny Squannit who is ill, and it is up to the medicine woman Nonner Martha—a historic figure who lived centuries later than Pocahontas—to heal her.[6]

In *Makiawisug: The Gift of the Little People*, we discover the manito of the land, an Algonquin Gaea. She is Turtle Island's analog of the Logres mentioned in the Gawain narrative and also shares narrative features, at least, with the Fisher King and Osiris. Squáuanit, or Granny Squannit as the Mohegan of today know her, is the *weroanskaa* of the *makiawisug,* very small beings who resemble humans in all but their size and their *powa,* their magic. Their name means "tiny moccasins"; they are shod in small flower petals. Their size, their ability to shape-shift into various other animals such as Wolf and Bird, and some other supernatural or magical powers let us know the kind of beings they are. As the story begins,

> *Long ago, this land of the Mohegans was a very different place. Good*
> > *Spirits guarded the place.*
> *Among those Good Spirits were those called Makiawisug.*
> *They are the Little People of the Woodlands.*
> *In those days, the Makiawisug were many.*[7]

But they were shy and appeared only at night. One night Martha Uncas heard a knock on her door. It was late at night and a terrible storm was raging in the woods where she had her cabin. She went to the door and, long story short, was informed by her small visitor that "someone is sick." Being a healer, she didn't hesitate, but grabbing her bag and shawl she followed the little man out into the woods. She left her granddaughter, Flying Bird, sleeping, but she knew that the child would be fine because she was a well-trained, sensible girl, well schooled in self-reliance, as well-raised rural people usually are.

Nonner Martha's young, small escort led her for some time, but finally they came to a small village, and she was invited into a house where a very old woman, tiny as those who anxiously watched over her, lay. A young woman who stood near the elder came over to the healer and introduced herself.

> "I am Ponemah and this is Granny Squannit." She gestured toward the old woman. "She must be made well."
> > Nonner Martha was stunned. Of course she had heard of the great Granny Squannit. She knew that storms were caused when Granny was in trouble. Some storms were created by Granny's fights with Moshup, the giant. But this storm had been caused by Granny's illness.[8]

Martha could see that the old woman was very close to death, and she busied herself with her implements—herbal infusions administered in a turtle-shell cup, sage smoke brushed in the air over the patient with a fan made of turkey feathers—and with performing a masked dance accompanied by the proper chant-prayers, incense, and feather fan. When, after many days, Granny Squannit was restored to health and it was time for Nonner Martha to return home, Granny Squannit showed her a pack basket filled with gifts. She said there was a special one that Ponemah had made for her. Reminding the medicine woman to make sure her people always left corn cake and berry baskets for the *makiawisug* in the forest, she sent her on her way.

We find an echo of Granny Squannit in Disney's version of the Pocahontas story—to wit, Grandmother Willow, who acts as wise elder and supernatural guide for the young Pocahontas. Perhaps Granny Squannit, who would have been seen as a powerful being in seventeenth-century Algonquia, would have been the divine being Pocahontas addressed in her prayers for help and guidance in the strange world of English Christianity into which she had fallen. Maybe Granny Squannit inspired the writer of the film script of *Pocahontas* to "invent" Grandmother Willow. And perhaps chipmunk and hummingbird are a modern film version of the Little People.

More to the point, it seems likely that Granny Squannit is a familiar version of Nikomis, Grandmother, who came from the sky and caused Turtle Island and all it is to come into being, who is associated with times of great transformation.

In my own tradition of the Keres (Laguna Pueblo) People, the Great Being, God, is identified as *Ts'its'sti-Nako, Woman Who Thinks* (or *Makes* in the magical sense of creating). She is familiarly known as Grandmother Spider, and in this guise she appears in tales geared to the ears of youngsters and the informal storytelling sessions of times before the advent of television.

Meanwhile, back in the implicate-order *manito aki* of the English, Logres, Sir Gawain has sent his plea to the Great Being, God, and blessed himself as was proper. At that moment the holy magic of his situation began to make itself known to his bemused sight:

> Now . . . he was aware in the wood of a dwelling within a moat, above a lawn, on a mound surrounded by many mighty trees that stood round the moat. 'Twas the fairest castle that ever a knight owned; built in a meadow with a park all about it, and a spiked palisade, closely driven, that enclosed the trees for more than two miles. . . . Then he pricked Gringalet with his golden spurs, and rode gaily towards the great gate, and came swiftly to the bridge end.

Similarly, perhaps, the medicine woman-cum-spy, riding the bow of Argall's ship, hair streaming behind her in the river's breeze, rode gaily toward the palisades of Jamestown, entreating Granny Squannit and her *powagan,* or guardian spirit, to find her safe haven, came swiftly to her journey's end.

For Gawain, "The bridge was drawn up and the gates close shut; the walls were strong and thick, so that they might fear no tempest." A similar sight greeted Matoaka, for in the years since, as Pocahontas the child, she had last visited there, the English had built a sturdy palisade of towering logs, making strong and thick walls surrounding the buildings. These included a church, several dwellings, and a glass foundry. The fort was gated and well guarded against attack from the Spanish navy, Powhatan Indians, or whoever might be about and meaning them harm. And while she had been there so many times in years past, it had been quite a while since she had last set foot within the gates, and a great deal of building and wall strengthening had gone on in the meantime. Nor were there many of the original hundred-plus men—the Ancients, as they were known—still there, so she was greeted by mostly strangers, craning to see what Argall had brought, wondering, no doubt, if his entry with his captive meant war or peace.

While few of those she had known remained, as many had died in the Starving Time, others had returned to England, but one friendly face greeted her: "Among the 'Boyes' with whom Pocahontas had so indecorously disported herself," one would have been on hand to greet her that day in 1613.⁹ Now himself an "Ancient," Samuel Collier had served as Smith's page the night of Pocahontas's masque. Collier had survived starvation, accidents, illness, brawls, and the tortures meted out by various governors during the colony's short history. He would outlive Matoaka by five years, dying in 1622 when a musket misfired, killing him.

Gawain too was to be met by strangers in circumstances that were almost familiar to him, for while the castle he approached had appeared somewhat suddenly, once within its walls he found a society much like his own. The one familiar face that greeted him went unrecognized, perhaps because he was "fair impressed" by the fort's splendor. For even from beyond its walls he could see an edifice that bespoke wealth, power, and impressive style.

Likely Pocahontas was similarly impressed at the new sights that greeted her, along with the even greater indications of English encroachment on Powhatan land she would soon be removed from. She may have been alarmed, but surely the sights that met her curious eyes would have intensified her determination to stay and influence the swiftly transforming psychic sphere she was encountering. It is not likely that the Powhatans or their allies were unaware of the movements of the English. Since early on, fighting had broken out between them, largely because the aliens, who had been welcomed into the *tsenacommacah* as participating members, spurned all its laws and customs, scorning and denying at every turn all the efforts the *tsenacommacah* made to reach out to them, to welcome them and ease their adjustment in their new homeland.

The fighting between the local people and the English had become so pervasive that diplomatic relations had been severed and the war had heated up. It was this troubling turn of events that sent Pocahontas into exile among the aliens. They were in a huge way her charge; the failure of these aliens to accept the lawful ways of the *tsenacommacah* was her responsibility, and upon her shoulders rode the destiny of her people and, even more, of the grand tradition that she in every way represented.

The ramshackle huts that the mostly doomed first contingent of 104 men and boys had lived in had been replaced by stronger houses, and the settlement, now a town, had extended inland from the swampy marshlands it originally occupied. In his report "To The Truly Honorable and Right Worthy Knight, Sir Thomas Smith, Governor of the East India, Muscovia, Northwest Passages, Somer Islands Companies, and Treasurer for the First Colony in Virginia," writing as secretary "in That Colony," Rafe (Ralph) Hamor observed with keen appreciation that

> this I can say by mine own experience that corn and garden ground, with which much labor (being when we first seated upon it a thick wood) we have cleared and impaled, is as fertile as any other we have had experience and trial of. The town itself by the care and providence of Sir Thomas Gates, who for the most part had his chiefest residence there, is reduced into a handsome form and hath in it two fair rows of houses, all of framed timber, two stories, and an upper garret, or corn loft, high, besides three large and substantial storehouses joined together in length some hundred and twenty foot, and in breadth forty. And this town hath been lately newly and strongly impaled, and a fair platform for ordnance in the west bulwark raised. There are also without this town in the island some very pleasant and beautiful houses, two blockhouses to observe and watch lest the Indians at any time should swim over the back river and come into the island, and certain other farm houses.[10]

One can only imagine the young woman's thoughts on these improvements. Where forest had stood, cornfields extended. Where cottages in many particulars resembling those of her own towns had stood were sturdy houses more closely akin to those that dotted the English countryside at the time. The palisades, made of thick tree branches or trunks, had been greatly extended, thus "impaling" a much larger area, and defensive fortifications had been sturdily erected and strategically placed. To some extent these were defenses against local people, wild animals, and the like; but the bulwark, which served as a seawall, also functioned to stave off attack by the Spanish, whom the English feared. They had, after all, been at war for some time with Spain, and their movements along the Atlantic Coast were to some extent a continuance of that decades-long struggle.

Did she gaze around, go still inside, listen for the soft voice of the *manito aki,* the world of the spirit or, in more contemporary terms, the implicate order? Did she watch the trees sway, study the survivors of the Starving Time, and intuit the horrors of "order" that accompanied the arrival of the investors' salvation in the person of Sir Thomas Dale and his forces? Perhaps she searched the faces of these strangers for any hint of the manito, alien or local, trying to discern any traces of her earlier work over the years since John Smith had been remade Nantaquod. Nantaquod (Smith) had subsequently become the enclave's (her charges') protector. The record indicates she thought he was dead. When soul walking, did she seek his spirit? Perhaps, not finding it, she thought those strangers from the sea birds even more alien than they seemed. Did she sleep, pace, or sit, in whatever accommodations Argall or Gates provided her, reaching out for the manito, conversing with her guardian spirits, revisiting *apowa,* her vision? Or did she discover herself alone, a condition so common to us modern people, and already becoming familiar among Europeans during her time.

Perhaps when she learned, later, during her instruction in Christianity, that the dying Jesus himself had lamented, "Oh, God, why have you left me," she felt profound recognition of his words, realizing they spoke exactly to her own plight. Surely she sensed that many changes had been wrought upon James Fort, symbolized by the eerie absence of most of those she had cavorted with in those days not so long before. She must have noticed that the bustling, strictly disciplined, virtual enslavement of the colonists hinted at a new world order taking hold in this brave new world. Perhaps she had been unprepared for the abysmal difference between the two orders. Perhaps she had not known how terrible would be the fall into the void. As the brilliant Canadian fantasy writer Charles deLint observes, "The spiritworld. . . is a lot closer than most people think. Open yourself up to it and it comes in close, so close it's like it's right at hand, no further away than what's out there on the other side of the window. . . . Dangerous place to visit, outside of a dream."[11]

DeLint seems to give voice to what Pocahontas, from the place of her exile, could have been thinking. That year, 1613, under the iron hand of Marshal Dale, the English version of *manito aki,* Logres, was a dangerous place indeed. For Matoaka as she entered the fort, the question was whether or not she was still wrapped in her *powa,* and if so was it sufficient to protect her and continue to guide her as she went about her mission. James Fort, barricaded behind walls made of large logs, was as dangerous to any who entered as the place where Gawain met the Green Knight on the appointed date.

Although things were looking comparatively good, the people, characterized in a scathing comment of Smith's as "the dregs of humanity," were fearful, anxious, and beaten down, having been brutally reminded of their status as serfs by the harsh measures Sir Thomas Dale had taken to whip them into line since his arrival. Marshall Dale, as his office was titled, and the somewhat more lackadaisical acting governor, Sir Thomas Gates, were in firm control by the time Argall brought Matoaka in.

Europe had been making its way from the old ways of the quasi-implicate world it mostly had inhabited until quite recently. Part of its turbulent shift from quasi-implicate to full-bore explicate order was expressed in the international fight over dominance in Europe as the modern nation-state evolved. Widening political power fueled by the newly emerging market economy was the major development in terms of the explicate order. More significantly, the parallel movement in the implicate order, which gave rise to the outward changes, was best mirrored in the struggle in Europe over control of religious power. This struggle, which encompassed the Protestant revolution, including the widening split between state-sanctioned and "free" church systems, also saw the expulsion of Islam from as much of Europe as possible, and the concurrent expulsion or persecution of Jews and European pagans. Indeed, it can be argued that the Europeans' intercontinental adventure—the conquest of the Americas, much of Africa, and the Pacific Islands—was an extension of the period's great struggle over ideological dominance.

The wars of ideology that raged across the Continent to the borders of the Middle East through the fifteenth and sixteenth centuries were evidence of far greater upheavals within the implicate order. Awareness of those "inner" upheavals enabled civilizations still connected to that order (as were most of those of Native America) to know "what time" it was, as some traditionals put it. The knowledge Native peoples had was possible because a strong connection to the implicate order yields a particular kind of data. Those who know how to download the information and analyze it accurately know within a reasonable degree of probability what's going on in the universe.

In his study of metaphor as myth and religion, *The Inner Reaches of Outer Space,* the great mythologist Joseph Campbell remarks that "a mythology is not an ideology. It is not something projected from the brain, but something experienced from the heart, from recognitions of identities behind or within the appearances of nature."[12] Campbell's comment establishes an important point about the kind of knowledge these people could access, and with such a high degree of accuracy.

The Great Transformation, as the worldwide tumult had been characterized in Native thought, was played out as the struggle, the continuance of the centuries-long wars between the Ottoman Empire and Europe. It entered the phase that directly affected the Algonquin and other Native civilizations about the time that Chris Columbus set sail west. He embarked just on the heels of the final expulsion of the Moors from southern Spain and the expulsion, burning, or forced conversion of both the Spanish Jews and the remaining Moors. The final expulsion occurred beginning in 1492, when Ferdinand and Isabella took Granada. Unsatisfied with reclaiming Spain for the church, the Spanish continued the struggle until 1609, when the last Moriscos, Moors who had converted to Christianity, were expelled. Nor, one might observe, has that long, long war yet seen an end.

Mercantilism, or preindustrial capitalism, rose concurrent with Protestantism, leading to the establishment of the American colonies. This in turn led to the establishment of the United States, said by historians to be the first purely Protestant nation in the world. The States also lay claim to the title of first democracy, and it can be argued that it is the first, and so far only, independent capitalist polity in the world.

WHEN *MANITO AKI* AND *LOGRES* COMBINE

All of which may seem a long way from the sights that greeted Pocahontas on her return to the infant alien nation growing on the marshy shores of the Powhatan River, unless we keep in mind that due as much to her service as agent of sacred (implicate-order generated) change, the *manito aki* and *Logres* began their merging. Matoaka may have been somewhat overwhelmed at the sights that greeted her, as much by their strangeness as by the absence of most of those she had known in her earlier visits. Fable has it that she still grieved for her lost John Smith, but whether she actually did or not, and if she did, what the reasons for her grief might have been, are lost to us, except in a passage John Smith himself recorded concerning his meeting with her in England. Pocahontas didn't write reports, at least not in a form that survives in university archives or reissued publications.

No more the thoughts of Pocahontas than the events chronicled by the *Pearl* poet concerning the journey of Sir Gawain from human territory to the realm of the spirit people, who are, at least now, known to us as "The Gentry," or "Faerie." My great-grandmother, Meta Atseye Gunn, used to tell stories about the old gods at Laguna. She'd use the terms she thought were the English equivalent, so Grandmother Spider was "a Fairy," and the *nowish* were "Little People." She had been educated at Carlisle Indian School, in Pennsylvania, where she had been acquainted with works of English writers such as Tennyson and Spenser. She particularly remembered Spenser's *Faerie Queene,* from which she drew a great store of information about the Anglo-American culture around us, and into which she had married. I must hasten to add that her husband, my great-grandfather, Kenneth Colin Campbell Gunn, was not of English but of Scots descent.

However, I doubt that such fine distinctions mattered much to Grandma Gunn, particularly when she was telling us, her mixed-blood great-grandchildren, stories that reflected her way of thinking about things in ways she thought we'd best understand.

It is my surmise that the young woman saw her journey into English-land in the same light that the legendary Sir Gawain saw his into equally alien realms. She was on a quest, and a large, somewhat daunting fort swallowed her, cloaking her even further from modern sensibility than Sir Gawain and his courtly cohort have been cloaked.

The young visitor was met by a courteous crowd, and was quickly made comfortable for the few days it seems she spent in Jamestown.

Courteously, Captain Argall welcomed aboard his unsuspecting guest and the *weroance* and *weroanskaa,* Japazaws and Paupauwiskey, his wife, of the Patawamack, whom Pocahontas had escorted aboard the English ship at Paupauwiskey's plea that Matoaka introduce her and Japazaws to the captain so they could go aboard and see an English ship when Argall was at anchor near Pastancie, a Patawamack town.

As Rafe Hamor, who at the time was the company's official recorder, or secretary, tells us, there "the best cheer that could be made was seasonably provided; to supper they went, merry on all hands, especially Iapazeus and his wife"[13] Soon enough things got strange. Soon enough Argall separated his charge from her companions so that he could pay them the price of their betrayal: a copper bowl and some less valuable items. Then the genial captain announced that the Princess Royal of the Powhatan nation was his captive. His announcement was met by much wailing on the part of the Patawamack leaders and the young woman who had rejoined them. Having settled his business with the local leaders, Captain Argall set sail with his captive, soon restored to her customary genial frame of mind. He reassured her that she would be well treated, as befitted a princess of her status, and, making good his word, settled her in the gunner's quarters, the best on board with the exception of the captain's own.

Upon his arrival in Jamestown, before dropping anchor, Argall sent a message to Wahunsenacawh regarding the captive Matoaka and the terms for her return to her people. In a later message the English assured Wahunsenacawh that Pocahontas would be "well used," as the king had asked, but, alas, as Wahunsenacawh had failed to meet all the terms set for her return, they would keep her until the terms were met. Then, "we should be friends."[14]

As promised, she was well accommodated during her stay in "James Towne." On her arrival she would have been dressed and coiffed as a respectable adult woman of the *tsenacommacah*. But Jamestown was only a stopping place on her journey, as the fine palace where Gawain passed a few days until New Year's was for him. Soon he left the friendly people and the feasting and dalliance that had beguiled him during his visit, and headed out into wilderness once again to find the Green Chapel.

Similarly, Matoaka was in Jamestown but a few days before Sir Thomas Dale, accompanied by the young divine the Reverend Alexander Whitaker, and John Rolfe, an English gentleman then serving as soldier to Dale, moved her upriver to Henrico. Immediately, Marshal Dale enlisted the aid of some of the settlement's recently arrived English women to see to it that Matoaka was properly covered, so as to spare the devout the sight of her bare bosom and tattooed arms and legs. Soon the "hostage" was bedecked in stiff and undoubtedly uncomfortable English women's garb. Upon her resettlement she came under the tutelage of Whitaker, a recently arrived Puritan minister who was a close associate of Dale's. The "wilde" girl who had visited and succored the fort was transformed into a conservatively garbed, dark-skinned English woman.

Matoaka had gone from Jamestown to the new settlement of Henrico, about ten miles south of present-day Richmond, Virginia, toward her meeting with the English magicians' training school. In his parallel narrative, Sir Gawain journeyed from the palace toward the Green Chapel to meet the Green Man, a magical being from the tradition that Matoaka earnestly desired to study. She had traveled by boat, upriver to her new home, while Gawain rode his faithful steed.

Then the knight spurred Gringalet, and rode down the path close in by a bank beside a grove. So he rode through the rough thicket, right into the dale, and there he halted, for it seemed him wild enough. No sign of a chapel could he see, but high and burnt banks on either side and rough rugged crags with great stones above. An ill-looking place he thought it.

Then he drew in his horse and looked around to seek the chapel, but he saw none and thought it strange. Then he saw as it were a mound on a level space of land by a bank beside the stream where it ran swiftly, the water bubbled within as if boiling. The knight turned his steed to the mound, and lighted down and tied the rein to the branch of a linden; and he turned to the mound and walked round it, questioning with himself what it might be. It had a hole at the end and at either side, and was overgrown with clumps of grass, and it was hollow within as an old cave or the crevice of a crag; he knew not what it might be.

"Ah," quoth Gawain, "can this be the Green Chapel? Here might the devil say his mattins at midnight! Now I wis there is wizardry here. 'Tis an ugly oratory, all overgrown with grass, and 'twould well beseem that fellow in green to say his devotions on devil's wise. Now feel I in five wits, 'tis the foul fiend himself who hath set me this tryst, to destroy me here! This is a chapel of mischance: ill-luck betide it, 'tis the cursedest kirk that ever I came in!"

Helmet on head and lance in hand, he came up to the rough dwelling, when he heard over the high hill beyond the brook, as it were in a bank, a wondrous fierce noise, that rang in the cliff as if it would cleave asunder. 'Twas as if one ground a scythe on a grindstone, it whirred and whetted like water on a mill-wheel and rushed and rang, terrible to hear.

"Stay," quoth one on the bank above his head, "and ye shall speedily have that which I promised ye."

In the end, Gawain underwent his magical trial, and more or less passed, an event that follows the pattern seen in the Sky Woman myth; both hint at the path Matoaka must walk if she is to succeed in her quest. The Green Knight didn't kill Gawain but spared him, because, when he told Gawain that the woman whose green girdle the knight wears so jauntily (which signified some intimate time spent with her) was his own wife's, Gawain admits he did a wrong. The young knight's honesty caused the huge being, guardian of Faerie, to spare him. But the fact that he had been tricked into his embarrassing failure is as significant a detail in his acceptance by the Faerie Man: those who walk in the world of the implicate, whether it's called *Logres* or *manito aki,* learn that trickery is part of the challenge. Successfully negotiating the perilous currents there requires a firm and steadfast adherence to one's own goals and the values that inspire them. We realize that the Green Knight was Gawain's all-too-gracious host, the proffered entertainments a deceitful ploy, that Gawain has been walking in England's *manito aki,* all unaware, until the giant Green Knight revealed his magical—or sacred—identity. As those who walk the implicate-order paths warn us: walking in the *manito aki* is dangerous, unless you do so in dreams.

So indeed with our Princess of the Woods. Her captors, seeming generous, respectful hosts and she their royal guest, were reading off a page taken from a book she had never heard. Given her training, one would think she would know that things don't go in any predictable, known way in Faerie. She might have known that whatever was unimaginable to even a trained Beloved Woman—young, to be sure, but a full matron after all—was the likeliest course that would open before her astounded eyes.

And indeed, if the changes to the fort since her last visit and its prospering extensions into the new Jamestown were any guide, she must have been at least surprised. After all, the changes had been wrought in only a few years. "*Manitto-manit!*" she may have exclaimed: "They are Gods." Later, after she had learned to read English, after she had left the strict mercies of the Reverend Whitaker and cast her lot with the forward-looking, if acolyte of *apook* (tobacco), John Rolfe, she discovered William Shakespeare.

Shakespeare, her near contemporary, was the immortal and as yet unbanned bard who had come to grace the London scene with his dramatizations of favorite tales and news alerts.[15] He had penned, as voiced by the character Miranda, in *The Tempest,*

> *O, wonder!*
> *How many goodly creatures are there here!*
> *How beauteous mankind is! O brave new world*
> *That has such people in't!* [16]

Shakespeare was imagining the new lands so recently revealing themselves to European eyes. Pocahontas, however, when she became acquainted with his words, would have interpreted them otherwise. Because her frame of reference was just about the obverse of Shakespeare's, she would have seen the New World as England, not the *tsenacommacah.* For England was as new to her as "Virginia" or Bermuda was to them.

Indeed, she might well have seen a performance of *The Tempest,* or so believe the researchers of the Jamestown Rediscovery Project. In volume 3 of their series Jamestown Rediscovery, the caption accompanying figure 7 identifies among its several images:

> William Shakespeare, a signet ring found at Jamestown in 1996, . . . and an artist's rendering of the wreck of the Jamestown supply ship, Sea Venture in Bermuda in 1609. The ring is engraved with the Strachey family heraldic crest (a displayed eagle with a cross on its breast). William Strachey's written account of the Sea Venture wreck inspired Shakespeare's famous play, *The Tempest.* [17]

In volume 2 of Jamestown Rediscovery, figure 11 bears a caption that reminds: "Elizabethan symbols suggest that despite the town's namesake, King James, the Jamestown settlers retained their Elizabethan roots."[18] The easy deceptiveness of the situation might have helped lull Pocahontas. After all, had she been suddenly transported to, say, the Chesapeake Bay area of the late twentieth century, she might have been put on guard, and no kindness, respectfulness, or smiling generosity of her captors would have deceived her. But the James Fort she entered was enough changed to amaze her, but not so changed that it would disorient her.

In another of the Jamestown Rediscovery volumes chronicling the excavation of the original James Fort and Jamestown that has been under way since 1994, project leaders note that, as artists' conceptions of James Fort based on eyewitness accounts from 1607–11 show when they are compared against an "eyewitness image of [an] Algonquin fortified village [around] 1585, . . . pre-industrial revolution European material culture and the Native American material of 17th century Virginia had much more in common than, unfortunately, either of them realized."[19]

I would add that the commonalities went quite a ways beyond material culture. They held in common a basic worldview that was changing in the seventeenth century only among middle-class English.

An Irish writer, Lord Dunsany, coined the phrase "beyond the fields we know" to describe fairyland:

> First there's the comfort of the fields we do know, the idea that it's familiar and friendly. Home. Then there's the otherness of what lies beyond them that so aptly describes what I imagine the alien topography of fairyland to be. The grass is always greener in the next field, the old saying goes. But perhaps it's more dangerous as well. No reason not to explore it, but it's worthwhile to keep in mind that one should perhaps take care.[20]

Perhaps this admonition might have stood Pocahontas in good stead had she heard it. As it was, seemingly undaunted by the "perfidy" of her hosts back at Pastancie, or by the failure of Wahunsenacawh and the Powhatan Council to redeem her, she let herself be coifed, garbed, and conditioned in a mode of great satisfaction to her captors, the alien others from far, far away.

In his conclusion about Matoaka, based on accounts by Strachey, Smith, and others concerning this period in Pocahontas's life, Edward

Haile speculates: "Pocahontas is now happy to throw her lot in with the English." In earlier years, he writes, "she was her father's [*sic*] spy and negotiator at John Smith's fort. A few [years later] she was an incarcerated captive." Here Haile appends a note that throws an interesting monkey wrench into the whole Pocahontas-as-innocent-who-is-captured-all-unawares scenario: "Or what was her status?" he asks. "I find it curious that in the small confines of Jamestown she goes unmentioned by another captive, Don Diego de Molina."

Unfortunately, without informing us further about his suspicions, Haile continues: "Now, always irresistible, she delights in the role of England's peace emissary to her father [*sic*]. She eagerly and quickly becomes a Christian lady and bewitches everybody in sight."[21]

We don't usually think of a woman as an adventurer, but Pocahontas was. Although cast as an ingenue, the love interest to John Smith in the founding-of-America tale, she was following a different narrative thread, one widespread in North America before the Europeans brought their narrative tradition over with them.

In some ways, we can unpack the story of Pocahontas as a tale of narrative collisions. Such an analysis makes for thoughtful reconsideration of the concept of the universality of certain story lines. The hero tradition comes to us via a very long narrative history that goes back into the mists of the earliest of the Indo-European cultures and that informs all texts from that tradition, whether sacred or folk, literary or popular. This tradition has followed Europeans to Hollywood, whence it has blazed across the big and small screens alike. Since its inception it has dogged academe like a faithful and ubiquitous companion. So familiar is this companion that it is all but unseen until you consider narrative conventions from other streams, both from North America and Euro-Asia. Among the Celts, for instance, there are two traditions.

But there is a parallel tradition—that of the women. It focuses strongly on the connections between the seen and the unseen realms, the world in which creatures other than man hold positions of authority and power, the world in which the Old Knowledge is woven like lace, or embroidered like tapestry, or decorated like pots, jewelry, and grooming and cooking implements. It is not dead by any means; it is thriving in the knit work of Scotland, Ireland, and Wales, where given patterns of sweaters designate village, clan, and family history. It is elegantly preserved in the fine lacework of ladies from northern Europe, the needlework of women of eastern Europe and lands to the north and east of it.

One can make a convincing argument for the origin of both medicine and magic in the woman's tradition; such an origin is recollected in stories as well as beadwork, birchbark paintings and weaving, pottery, metalwork, and house and village design, in herbology, horticulture, and the traditional dress of the tribes. It is traced not only in clothing worn by women but also in the clothes worn by men who have been dressed by their women and combed and ornamented as though they were living, breathing Ken dolls. It was often a woman who dressed a man, whether he was her husband, brother, father, uncle, or grandfather, not because she was that man's slave, but because Native women took a lot of pride in having the best-dressed men in the group parading their womanly aesthetic and arcane skills.

It is certain that the first tools were the digging stick and the scraper, the earliest crucible the cooking pot, the first transformation ritual the making and keeping of the fire, which our stories remind us was brought to us by a wise and all but invisible female adventurer, Spider Grandmother or her analog.

This is not to say that women did it all, but rather that two narrative traditions, two histories, two sets of facts—the male and the female—ride together, side by side, throughout human history, and that while the one we are the most familiar with is our male-centered narrative paradigm, there is a female-centered narrative paradigm that complements it. In the traditions, which is of greater consequence depends on who is narrating and which audience is involved, although the narrator is the more significant of the two. As the northwest Indians' wisdom holds: it takes both male and female to hold up the sky. A modern might add that it takes positive and negative poles to run a current or keep a planet in existence.

But in a traditional native story about a female, whether god, supernatural being, or human woman, the female tradition usually holds sway, although there are many examples of parallel narratives in which the figure may be male or female but the events narrated are much the same. In all I have seen that is written about Matoaka, Pocahontas, nothing at all is narrated as if it existed within the female tradition. In the woman's narrative tradition, one of the cycles that occupies a prominent place in most traditions is the one identified as the "star husband" motif by folklorists.

One of the best women's stories is from a western Algonquin tradition, the Ojibwaj, and it is about two semi-supernatural women, sisters named Oshkikwe and Matchikwewis.[22] I am deliberately framing this in informal terms to signify that these stories are as relevant to ordinary people today as they ever were.

Once upon a time, or maybe last summer, Matchikwewis and Oshkikwe were lying on the grass on a hill above their village. It was late summer, and the sky was bright with falling stars. They watched in awe for a time, trading comments and jokes now and then, until it was very late. Then one of them mentioned that one of the stars was a rather good-looking dude. Her sister agreed, but said she was more taken by one of his neighbors, who looked more trustworthy to her. "Oh, you would. You're so staid and proper," joked the more adventurous sister, Matchikwewis. "He's old!"

It was an old conflict between them—not the kind that led to ill will, but the amiable, joking kind. Their differences yielded as much a bond between them as a point of teasing. This was probably because, like most of their people, they loved to laugh, and enjoyed poking at each other's characteristics. They often made up names for one or another of the community because the renamed had a pronounced characteristic, or had done something out of character and amusing, or had been caught out in something embarrassing. It was their way.

But on this particular night they were more interested in figuring out how to entice the star men to come to the hill where they lay. Neither had been courted for some time, not by anyone new and glamorous, and they were both chafing a bit at the ordinariness of their love lives. Laughing, they agreed that an unknown "teepee creeper" would definitely spice up their lives, which could get boring in such a small town. They spent some time picking over the relative good and bad points of the star men they each had an eye on, and eventually dropped off to sleep. Oshkikwe, always the more practical one, pointed to one star. "He looks good to me," she said. "He could put his moccasins under my bench anytime."

Matchikwewis didn't find her sister's choice at all interesting. "Yuck, he's old!" she exclaimed. "He looks *b-o-r-i-n-g.*" Pointing to a bright star directly above them, she said, "Now, he's more my type." So, comparing one star with another and teasing each other about their taste in men, they fell asleep.

Just before dawn, when the sky in the east was hinting at lightening up, Oshkikwe awoke to see two figures—men, she decided—sitting near the women. At first she wasn't sure what she was seeing, because they were leaning against a large rock that stood out against the top of the hill. One was reclining somewhat, resting on his elbow. Slowly she reached out her hand and nudged her sleeping sister. "Yo, Sis," she said quietly. "We have company."

Matchikwewis opened her eyes and raised her head so that she could see over her sister's torso and looked in the direction of her sister's gaze. "Uh, I see," she said. Or maybe she said "*Manitoo-manit!*" or the Ojibwaj equivalent. Now, these women were beyond giggling girlhood. They were full-grown women. So rather than ducking their heads and covering their giggles with hand held over mouth as proper young women would, they sat up. "Are you here?" said Matchikwewis, who was always the bolder of the two. It was a common way of greeting someone politely among their people.

"What makes you think that?" joked one of the men. Matchikwewis was thrilled to hear his deep rumbling tones. *All right,* she thought gleefully, but of course she kept her face calmly composed.

"You called?" said the other man. Then he added in a softer tone, "Well, maybe not. Did they call?" he asked his companion. "I heard somebody evoking us," said the deep voice. "But maybe it came from the village down there." He gestured with his lips and chin toward the village below that was just becoming visible.

The women glanced at each other. "Live ones," Oshkikwe murmured, stifling her own delighted grin. "Maybe," she replied, "but maybe it was us."

Soon enough the two couples had enough small talk. The women rose, sauntered over to the men, Matchikwewis swaying slightly like women do at Squaw Dance, Oshkikwe's gait more sedate. They went off with their new guys and soon were settled in the men's village, high above the clouds. There they remained for a long time. Oshkikwe got pregnant, then had a boy, and this occupied them for some time, as did getting acquainted with the strangers who lived in the sky, and with their ways.

But after a long time in the star nation's world, they began to get homesick, and talked more and more about returning home. No one seemed to be taking the hint, though. The star people went about their business as though the two had always been part of their community, until the sisters began to grow very discouraged. Sometimes they would gaze on their old village that lay far below; on the river that made its slow way from the huge lake, past the town, and on into the blue distance; on the tiny people coming and going. They would try to make out the individuals they saw, telling one another that this one was surely so-and-so, just look at the way he strode around, and that must be so-and-so, she always moved like she was carrying a huge pack of stones on her back, and so on.

But they grew more and more depressed, losing even their interest in talking to each other. They seldom looked down, because it made them feel too bad.

One day they were sitting, disconsolate, watching the boy toddling about. They were weaving baskets, but neither had much enthusiasm for their task. An aged woman came up to them and sat. In her arms she was carrying a large, beautifully made pack. It was finely woven and capable of expanding, as they knew. They had seen enough strange things during their sojourn that they paid small attention to the basket, until the woman began to speak.

"I have noticed that you have been yearning," she said, eyes lowered on the needlework she had taken from her belt pack when she sat. "I think there might be a way to take care of this," she added, giving the needle a tug.

The sisters listened carefully, although, being polite, they didn't look at her but watched the boy. They kept their own hands occupied with the bit of basketry they had been working on when she joined them.

Indicating the pack with her pursed lips, the elder said, "This one can get you there. You take it outside and over to the edge. There's a cord attached to it that you can attach to a tree or a good solid rock. Then you get in with the baby and pull the flap closed. Think about your home. About your village, how it sounds and feels and smells. How the breeze is along the river, the sound the grass and the bushes make talking to each other. The birds. Remember the whitefish, and your mother."

The women worked on their projects in silence for a time. Then the baby crawled over to them and the elder reached out to smooth his hair off his forehead. She reached into her belt pocket, drew out a sweet, and handed it to him, regarding him steadily with her clear gaze.

The boy began to suck the sweet and soon crawled over to his mother and fell asleep with his head in her lap. The women chatted a bit quietly, and after a considerable time had passed the young women unfolded themselves, gathered their things and, stowing them into the beautiful pack the old woman had indicated, moved away. Oshkikwe carried the boy on her back, secured with her shawl.

That night they went out from their sleeping quarters and made their way to the edge. Following the old woman's instructions, they gathered themselves into the capacious pack, closed the flap, and settled in.

In time they felt the pack begin to descend. Concentrating on their home as they had been told, they waited. Soon enough they felt themselves settling onto firm ground. Opening the flap they looked out. They were home.

ONE THING FOLDS INTO ANOTHER

In his discussion of the implicate and explicate orders, David Bohm describes the former in terms that remind me of what I call the chiffon cake theory of universal order, in which when one folds the beaten egg whites into the batter gently one re-creates the process—and, I'm told, the mathematics—of the universe in action. Or, as Bohm describes it:

> In terms of the implicate order one may say that everything is enfolded into everything. This contrasts with the explicate order now dominant in physics in which things are unfolded in the sense that each thing lies only in its own particular region of space (and time) and outside the regions belonging to other things. . . . What distinguishes the explicate order [is] a set of recurrent and relatively stable elements that are outside of each other.

I call it the chiffon cake theory because the process of folding egg whites into batter is a graphic and dynamic illustration of the principal Bohm is describing; the events he describes are experiences, not abstractions. Shamans and medicine people experience the fluidity of the implicate order as a fact of life, and the fixedness of the explicate as a symbol, an abstraction. Bohm continues, in part: "In the prevailing mechanistic approach . . . these elements are taken as constituting the basic reality."[23] And here we once again stumble on the crux of the matter: Pocahontas lived in a world where the implicit order dominated thought and therefore all social forms. The shocking thing, for her, might have been the relocation to an order that had departed from that paradigm—although not to the extent we have—and that was rapidly growing accustomed to thinking in terms of the explicate mode: fixed objects in a dead field. Some might see these differences as different brain functions based in the structure of the brain and which differ between male and female subjects.

Recent studies in human brain function demonstrate pretty decisively that women process input in a right-brain-dominant way, although both hemispheres are engaged, as electronic "blips" on readout screens indicate. In contrast, men's brains process input in a left-brain—that is, mechanistic, fragmented, and compartmentalized—manner. The research and its implications are highly suggestive: they provide a powerful way of exploring the differing dynamics of Western and indigenous civilizations that is culture free but capable of being descriptive and politically unloaded. Perhaps in this way such investigations and applications can provide a fresh point of view that can be at once accurate and connecting about "cultural" differences, their reasons, and their development.

Those more politically inclined use the bilateral brain-function model to explain how the present world functions, and from there to assign culture-laden values to these operations. The left-brain function is thus labeled "patriarchal," while the function of the right brain is seen by both pro- and anti-feminist camps as "feminine." The analytic function assigned to the left brain is the standard by which moderns interpret—that is, make sense of—their world, while the right side, characterized as intuitive, artistic, creative, is where boundaries and differences are dissolved. Thus, the rational is equated with the light, fire, and day, while the other, the "irrational," is equated with shadows, water, and night. Those same "irrational" qualities were attributed to Pocahontas's people, who thus were categorized as evil and were associated with the devil, while their way of life was consigned to the wilderness. Both Indians and wilderness were seen as given to civilized men to tame, making the Virginians' success with Pocahontas the signal event they took it to be.

The world of the Powhatan, the Dream-Vision People, can be profitably reflected upon from the vantage point of Bohm's approach. It is a process that Peat has applied, with some success, to aspects of Blackfoot thought and tradition, and his exploration forms a model for our own. But after considering the bicameral brain model the discussion can pursue a further avenue of inquiry: if explicate thought is "mechanistic," it is also by definition "rational." Basically, it operates on the principle of the old dog: if you can't eat it or have sex with it, pee on it. Worse: if you can't see it and measure it, it doesn't exist; should you believe it does, you're a candidate for professional attention. In Pocahontas's case that professional attention was a Puritan educator and spiritual mentor. In ours it might be a minister or priest; more commonly it is a mental-health specialist. This belief system has its roots in seventeenth-century thought, religious and philosophical, including medical, and has persisted over the ensuing centuries.

The currency of this four centuries' belief that the nonrational is nonexistent, despite the best efforts of physicists and other science theorists to demonstrate otherwise throughout much of the twentieth century, is reflected in Helen Rountree's contemporary work. Rountree's explanation of Powhatan beliefs reflects the beliefs of John Smith and his contemporaries, the "new men" of Pocahontas's time. When Rountree informs us that the worship of images was rampant among the Algonquins of the period, she hastens to assure readers that the English *knew* they were bogus. "The English knew, as Powhatan laymen did not, that Okeus' image [figures made of wood and leather that held the god Okeus, or *Oke's powa*] was that and nothing more, for within five years of founding Jamestown, they raided several Indian towns and 'Ransacked their Temples.' "[24] The spin Rountree puts on the contrasting systems of belief is evident: the Powhatans have false ideas, while the English *know* the facts of the matter. There is little reason to suppose that the *okees* the Indians brandished were "images" and nothing more. Perhaps in the eyes of some of the foreigners who were some of the world's earliest disbelievers in nonmaterial realities that was so. But two different points of view are being offered in the event, rather than a false belief being faced with factuality.

There are number of reasons to explore the possibility that different experiences of reality were in fact occurring, in what I have elsewhere defined as "gynosophic" traditions, by which I mean social systems organized along the lines of "feminine" consciousness as separate from those experiences of reality experienced in those traditions identifiable as masculinist.[25] It seems to me that characterizing these systems as "implicate order" systems might help dispel some of the confusion that necessarily arises when gender issues are injected into the analysis, particularly because in modern minds "feminine" implies the unconscious, emotional, and nonrational.

What remains clear is that the two systems do indeed differ widely in organizational modes, and these modes result in a particular segment of the overall community being privileged over the others. This fragmenting of a whole state system results in most of the ills that currently plague our human world.

In implicate-order civilizations, a number of parameters can unfailingly be observed: the prosperity of the individual is inextricably dependent on and answerable to the prosperity of the group; the clan as a whole, rather than the individual man—or woman—is always privileged over "nuclear family" (which in itself is, literally, an unimaginable concept), while the clan itself is seen by all involved as one of a number of units within the entire system. This whole-system outlook means that there is a clear perception of relationship among all the beings that co-inhabit the *manito aki*. It means that in such systems, balance and harmony are seen as laws of nature, and such virtues as generosity and integrity are held to be ways in which humans act when they are in harmony with the whole, walking in balance.

In these systems, tolerance of differences among different kinds of people, be they European, Siouan, or Algonquin, human, animal, vegetable, mineral, climatological, or astronomical, is a prime rule. In these systems it is commonly understood that all beings have their prime movers, their gods, their relation to the all that is, and one does not interfere with that relationship

One unfailing sign of an implicit-order social system is what in present-day discourse is called matriarchy. That is, when a system is matrilineal, matrilocal, and matrifocal, it is an implicate-order system. This is not to be confused with female dominance. While "matriarchy" as a system has been theorized as the contrary to "patriarchy," no system of female dominance has been documented in either hemisphere. Those systems dubbed "mother right" systems—that is, systems that are matrilineal, matrifocal, and matrilocal—are actually misnamed. The idea of motherhood, so profound in Western consciousness, is not as central in the Native tradition. The clan, or, in some cases, the band, is the basic, fundamental social unit. The importance of the feminine is not the capacity of women to bear children; it is the fact that the feminine is itself more closely reflective of natural law and the "invisible" system that informs that law. A system based in feminine consciousness must, of necessity, be inclusive—not because women have babies, but because the female brain is so constructed that hemispheric walling off of emotion from both idea and ongoing experience is not possible, rendering perception and expression, conceptualization and emotion a dynamic, interactive, ongoing, ever moving process. In brief, women can't help being focused on relationship and process; it's not how women will or decide to think; it's how women *do* think.

As we attempt to see our way through the explicate-order-based thicket that has grown densely over contemporary understanding of seventeenth-century indigenous America, it is necessary that we keep in mind that the reporters at the time were of the explicate-order persuasion. Complicating matters further, with each decade the researchers, analysts, and interpreters who follow the records that earlier reporters left have grown ever more mechanistic and abstract; increasingly split off from whole-system thinking.

Respectful and highly skilled scholars such as Helen Rountree and Frances Mossiker, who are impeccable in their documentation and solid academic grounding, unfortunately miss the basic worldview that characterizes Native thought and that distinguishes it from scholarly and popular modern thought. Both scholars are contemporary women steeped in the thought system of the explicate order of modern times. Neither could pursue her academic career with the kind of disciplinary respect each has garnered and not be steeped in late twentieth-century explicate-think.

Thus, Frances Mossiker wonders why the Powhatans sent food to the starving English during a time when the colony, which the Indians had decided posed a serious threat, was close to extinction. However, to the local people, allowing people to starve to death is wrong; winning through honest combat, as they defined honest combat, was justified when one was under attack. Not only did they live within a system of honor that we today think of as right and proper—and very modern—but they further believed that acting dishonorably would inevitably result in the loss of one's place in the community, both explicate and implicate, village and *manito aki*.

Helen Rountree is unquestioning when her researches lead her to find that the English, reasonably enough in her opinion, see that the "*Oke*" or "*okee*" effigies the Native people use are constructions: wood and special effects, while the Native people superstitiously believe them to be real spirits or gods. The English soldiers comment on the Indians' superstitious natures and recount the steps that they, the soldiers, took to prove to the Indian fighters that their *okees* were nothing at all. They would take what they called the "effigies" away from fighters and burn them. As they manifestly survived the act, they supposed they had made their point. Rountree doesn't mention that this action was being taken by men who believed that witches who could levitate were actual beings, and that by burning those accused of being witches one could wipe out the power they wielded.

Neither scholar would have been likely to have made this mistake had she been trained to recognize implicate-order systems and the oddnesses—events that seem bizarre in an explicate-order context—that can, and must, occur within them. As the noted researcher and scholar of brain function Edward de Bono observes, the brain can see only what it is prepared to see. The structure of academic thought is not prepared to see implicate-order data or patterns. The world of the *manito aki* is a world where human and "other" share space. In such a world, divisions between "natural" and "supernatural" don't exist, and behaving as if they did becomes psychotic. Within its parameters one does not let others starve. Enemies are to be honorably confronted and, if necessary, vanquished via bloodshed. But even when condemned by their own shiftlessness, they are not left bereft of necessities. Similarly, according to the *okee*-burning English, the Indian fighters expressed great dismay when the English defaced an *okee*. The English reported that the fighters were dismayed at seeing their belief revealed as superstition; the soldiers were unaware that the grief may have been an expression of horror at sacrilege, not grief at having their belief revealed as false.

How different anthropological and historical studies will be "[w]hen one works in terms of the implicate order, [and] begins with the undivided wholeness of the universe, and [sees that the task] . . . is to derive the parts through abstraction from the whole."[26]

"How Carefully were they to instruct her in Christianity," wrote John Smith in *Generall Historie*. He was referring to the efforts of Thomas Dale, Alexander Whitaker, and John Rolfe.[27] We are reminded how Pocahontas was initiated into the order of the Christian lodge. While it was never seen that way by her instructors, it was the most likely interpretation Pocahontas would have put on her instructions.

When Matoaka went over to the English, perhaps it was as a dusky Mata Hari, clad in a buckskin apron and well tattooed. She must have understood implicitly, if you will, the totality of the commitment she was making. There was no way she, a granddaughter-devotee of Sky Woman and a vast and implicate-ordered tradition, was going to take a bit here and a bit there. When she went, she went all the way, eagerly, arduously, without reservation. There was, for her, no other way. Engaged in the intricate plot of world-changing affairs, she was far more than an agent for the Powhatans; she was the mother of a new race, as yet unborn. John Smith was, in some implicate way, the first and perhaps only child she had, as such things are counted in the *manito aki,* the world of the implicate order (in at least one Algonquin dialect). She took upon herself the entirety of the tradition she entered because that was the way of honor, a point she would make vehemently to John Smith when they met in England, scolding him harshly for his failure to do the same. When her meeting with the faithless Smith is considered at greater length in chapter 4, it will be clear that for her such was the honorable way. Thus, when Amonute-Matoaka became Lady Rebecca to her English contemporaries and Pocahontas for posterity, she entered the explicate order and, so far as we know, never looked back.

Master Dale had hurriedly removed her from Jamestown and the control of the lieutenant governor, Sir Thomas Gates, and his hand-picked divine, the Reverend Richard Buke, both capable men by all accounts. Some have wondered why she was not left in Jamestown under the care and guidance of these worthies, and various sugges-tions are made, mostly having to do with her putative father. How-ever, removing her from the well-defended James Fort and taking her far upriver to Marshal Dale's recently built Henrico (named after Prince Henry, son of the king, so that the Dale contingent's loyalty to the monarch would be unquestionable) might seem counterindicated. Perhaps the reasons are not so obscure.

Marshal Dale was an ambitious man, and no fool. His hand-picked young Reverend Whitaker, in charge of Pocahontas's religious indoctrination, was a Cambridge man, while Gates and the Reverend Buke were men of Oxford, that ancient center of learning since it originated in the Roman ecclesiastic scholarship of the middle ages. St. John's College, Cambridge, was a center for development of En-glish Puritan thought, which held that the monarch should not be the head of the church, nor should the state in any way have authority over it. The young Reverend Alexander Whitaker was a graduate of St. John's College, Cambridge, where his father was Master and Regius Professor of Divinity. The Cambridge group, the most dedi-cated to the Puritan agenda, wanted decisive power within the Vir-ginia Company. The company had been originally funded by the City of London. London, of course, was the seat of both royalty and the still young and unformed Church of England. Oxford University was closely aligned with both church and monarch, and had been since its founding in early medieval times. Cambridge, on the other hand, was not so ancient nor was it so enmeshed in the power hierarchy. Thus those who hailed from Cambridge were jockeying for power in a rapidly changing and greatly prospering world economy. That is, our men from Cambridge wanted a seat at the table of power, a place at the uncommon trough.

At the time, it was essential that England establish a strong presence in the New World to counter the empire-building ambitions of both Spain, the reigning world power of the time, and of England's historic rival, France. Elizabeth, the Virgin Queen—after whom Sir Walter Ralegh, with the queen's assent, named the English portion of the New World, mainland branch, "Virginia"—defeated the Spanish armada, put an end to English religious warring, and presided over half a century of peace and increasing prosperity. The City of London had enormous capital investment and a fiercely defended tradition of primacy within the country, controlling England's international trade as well as its intellectual and political life. The upper classes there, vying for a seat at the table of power with the growing merchant-gentry, were firmly attached to the rule of privilege. This included the local control, via the palace, of the church. As the Reformation movement was gathering steam in those change-filled, tumultuous years, these families and the major guilds, all of whom were invested in the City of London's portion of the Virginia enterprise, wielded every influence in their grasp to maintain that primacy.

Elizabeth I's father, Henry VIII (1509–1547), had broken with Rome and the pontiff over kingly divorce rights. In the years after the accession of Elizabeth, the influence of strongly anti-Papist, politically active merchant enterprises, fueled largely by the forces of the Continental Reformation movement, had increased in England. Elizabeth's successor, James I, had a situation on his hands. Under Elizabeth the overt bloodshed between warring religious factions had stopped. Elizabeth was a Protestant, but of an undetermined sort; in her day it was sufficient to be either Catholic—that is, Papist—or anti-Papist—that is, "Protestant." Under Elizabeth (1558–1603), the church in England returned to the Protestant fold after a brief and bloody interim of reestablished Catholicism. The Elizabethan Settlement decreed outward conformity to the official church, the one under the auspices of the monarch; but no policing of private belief was ordered. For the most part, this church resembled in many particulars the Catholic rites of the previous ten centuries, but the supremacy of the pope in Rome was replaced by that of the monarch, under whose authority the Archbishop of Canterbury was the primary church head.

In the eyes of the Anglican Church, the apostolic succession was not broken with Henry VIII's Act of Supremacy (1534). The Archbishop of Canterbury now held his high office under the authority of the monarch, which in the early seventeenth century was Elizabeth, followed in 1603 by James I. Such royal authority is still the case today in that country, despite the Reformation and the religious civil war that raged in the mid–seventeenth century under Oliver Cromwell, "Lord Protector." The civil war was a direct consequence of James's confused stance on Protestantism, coupled with rising sentiment in England to follow Calvinist doctrine in every particular, including the separation of church and state. What the straitlaced Calvinist Puritans saw as dissolute behavior among James's courtiers did little to smooth their growing ire over decisions the king made that, in their eyes, favored Catholics too much.

The no-nonsense Dale, perhaps seeking to establish his primacy, set to having a town constructed far from the somewhat less brutal Sir Thomas Gates's bailiwick, Jamestown. Henrico, up the James River around eighty miles—about ten miles south of present-day Richmond—served to highlight the political distance between the marshal of Virginia, Dale, and the governor of Jamestown, Gates. Taking possession of the most precious item so far cast up from the *tsenacommacah,* Pocahontas, Dale hurried to Henrico his royal captive, who seems to have thought of herself as his honored guest. Perhaps she viewed him from the frame of her own agenda, which would make him the Powhatan equivalent of her knight on the chessboard of the Great Change.

There are some interesting coincidences that occur in the Pocahontas saga; these coincidences are perhaps synchronicity, perhaps serendipity. For example, there is the coincidence in the names of those connected with St. John's College, Cambridge, in her sacred mission. The coincidental pattern forms at the establishment of the great university in 1511, about seventy-five years before her birth, and finds its full expression during the years near her death:

> St. John's College was founded in 1511, by the executors of Margaret, Countess of Richmond and Derby: . . . This being a divinity college, the fellows are obliged to be in priest's orders within six years from the degree of M.A., except four, who are allowed by the master and seniors to practise law and physic: the electors are the master and eight senior resident fellows, in whom is vested the entire management of the college concerns.[28]

Among the worthies listed in the roster, two, in addition to Dr. William Whitaker, play a role in Pocahontas's life: the poet and critic Ben Jonson, and the alchemist, mathematician, and Court astrologer Dr. John Dee.[29] Dee was a significant force in England's connection to the "other side"; he was famous for his Enochian magic, a system that involved contact with angels. Enoch was one of the earliest of the prophets in the Old Testament. Dee was part of a circle of metaphysicians and alchemists that included Sir Francis Bacon, Sir Walter Ralegh, Queen Elizabeth I, and possibly William Shakespeare. The other man, the court dramatist and famous poet Ben Jonson, we will meet at greater length in chapter 5, in which our heroine goes to visit the king a few years after her conversion/remaking.

Alexander Whitaker was the son of an eminent member of the Cambridge faculty, Dr. Whitaker, who was one of the foremost Puritan divines. His presence at the colony as Dale's chosen minister sheds considerable light on the political infighting swirling around the Virginia Company. On the one hand it was a royal enterprise, funded by private investors to be sure. But should it fall under Puritan dominance, it would be an investment opportunity, sans royal authorization. In its turn, this move to make the Virginia Colony and the company that backed it a public offering on the European market would make it the ideal location for the realization of the Puritan

dream: a place where the church would function outside the authority of the monarch.

The long-term results of the power struggles going on in England
and their allied consequences in Virginia netted the forces of Puritanism the firm beginnings of their own nation, perhaps much sooner
than Dale and his Cambridge society Puritan mavens might have
thought. The charters were revoked and the Virginia Company was
dissolved by King James in 1624. Earlier, in 1618, the "company of
adventurers and planters of the City of London" established the First
Colony in Virginia and, in 1624, the foundation of private property
in America was formally set; just before that world-changing event,
the newly appointed governor of Virginia, Captain George Yeardley,
was "directed to further the colony's welfare 'by settling there a laud-
able form of government.' The Virginia General Assembly, America's
first representative legislature, was elected and convened."[30]

In terms of the situation Pocahontas entered when she went
among the English, the spirals of the receding implicate order that had
still percolated through Elizabethan thought were fast fading in the
growing glare of Protestant rationalism. By the early seventeenth century, the iron imprint of the explicate order was well begun.

Sir Gawain, a figure from the English and Continental Oral Traditions a thousand years before the seventeenth century, journeyed
from the early medieval equivalent of explicate order to full-force implicate order to meet his destiny. Pocahontas, for her part, was on the
opposite trajectory. She moved out of the implicate-order system of
the *tsenacommacah*—where, to be sure, change was afoot—into an explicate order similar to the one Gawain left, albeit one a thousand
years further along toward full mechanistic determinism than his
Bronze Age Camelot had been.

FOREST SHADOW TURNS INTO BRONZE

So Pocahontas has been abducted, or has colluded in her abduction—
or even orchestrated it. We have explored the parts played in implicate
and explicate space in this shadowy narrative. The collusion of narrative traditions, which also means worldviews, and its consequences is

moving the narrative from its never-ending, never-beginning sources in the Algonquin sacred Oral Tradition into the conflict-centered mode of the also emerging new world order. In a word, Matoaka is remade, and with her, the *manitowinini,* the people of the manito world, *manito aki.* Does the *manito aki* also transform, or does it continue along its own ancient independent trajectory? Similarly, although Matoaka becomes Rebecca, does the true Beloved Woman, member of the ancient Great Medicine Lodge, Amonute, undergo the change? One may well ask, because the "secret" name she confessed to those who initiated her into the ways of Christian civilization, English style, was only her clan name, Matoaka. She never disclosed her secret, sacred name, Amonute, to the English. It was uncovered almost by "accident" by Robert Beverly, who pursued inquiries into the earlier period during his visit to the colonies in the early eighteenth century. One of the few Powhatan men remaining in the tidewater area mentioned it to him, and he duly noted it.

Before we can begin to sort that complex issue, let's see how the remaking of the Powhatan princess was done by examining the rituals employed by her remakers: the Calvinist Cambridge-based Sir Thomas Dale, marshal—that is, acting governor—of Virginia; and his handpicked minister, the twenty-six-year-old Reverend Alexander Whitaker, son of the Cambridge savant Whitaker.

Wahunsenacawh is described by his chroniclers from across the sea as a savage leader, possessed of many wives, servants, and children, whom he takes from their mothers in infancy. He is bloody, frightening, and altogether barbaric, or so he appears in John Smith's chronicles. What then are we to make of Master Dale?

Mossiker describes him as a "military man and a stern disciplinarian." That's a bit of an understatement, given the measures he takes to restore discipline to the failing Virginia Colony. Smith had been a stern disciplinarian when he held the office of the president of the council, and believed that those he had chastised, beaten, whipped, or otherwise severely dealt with for "laziness" or other transgressions deserved such treatment. But Dale was an even more stern ruler: "Death was the penalty for lèse-majesté (disrespect to the sovereign or the sovereign's representatives)."[31] Thieves fared scarcely better, and in some cases one imagines that the miscreant wished himself seen as merely disrespectful of authority. Some convicted of theft only had ears lopped off or a hand branded. "And some which robbed the store, he caused them to be bound fast unto trees and so starved them to death." Should a man decide to "get the heck out of Dodge," the penalty was equally horrific: "Some he appointed to be hanged. Some burned. Some to be broken upon wheels. Others to be staked, and some to be shot to death."[32] The Spanish Inquisition, still active in Roman Catholic countries in that century, had nothing on Master Dale, and the first slaves brought to the colonies decades after his reign of terror fared no worse. However, we are assured that Sir Thomas Dale showed only the utmost kindness to his captive princess, at least publicly. Mossiker informs us that relations remained cordial between them until Dale left the Virginia Colony for more daring adventures across the Pacific.

When he removed her to Henrico, Dale consigned her to the "tender" care of the Reverend Whitaker, with the aid of the evidently even more tender John Rolfe, who may have hailed from Heachem, upcountry from Cambridge. Rolfe's "birthdate cannot be positively established, any more than can his origins. His ancestry has been the subject of endless research, speculation, and conjecture: generations of historians and genealogists have sought and failed to make positive identification of this particular John Rolfe with any of the several branches of the Rolfe family in England."[33]

The Rolfes of Heachem, in East Anglia, northeast of Cambridge, were a prosperous gentry family who, for lack of solid evidence to the contrary, have acquired the position of John's ancestors. Whoever the mysterious John Rolfe's family, he and the impeccably pedigreed Pastor Whitaker speedily and flawlessly turned Pocahontas from a mischievous wanton into a classic example of a fine Christian lady. She was, indeed, the peerless example of proper heathen conversion, and was for a very long time considered the most perfect and perhaps only perfect convert this heathen land ever produced.

Accounts of events surrounding the courtship and marriage of the woman now named Rebecca and John Rolfe are inconsistent and puzzling. For one, there is the matter of the identity of John Rolfe. Additionally, the odd way he notifies Marshal Dale of his desire to marry Lady Rebecca raises questions, even among those who knew him at the time. Also puzzling is the possibility that Pocahontas had married a Native man named Kocoum, a possibility questioned by biographers but not firmly settled.

While most accounts affirm the love between the Rebecca/ Matoaka (Pocahontas) and John Rolfe, particularly on his part, there is little evidence to support the tale other than Rolfe's odd missives to that effect, and they are contradictory. Did he marry the girl he loved, or did he just take a squaw because, presumably, his English wife had died while the ship they were on was becalmed near Bermuda and he needed a servant in bed as well as for kitchen, garden, and assorted duties? Or perhaps he married the prize of the ambitious Puritan faction because it was his Christian duty, as he himself declared. Additionally, there was the motive provided by Dale's and his London-Cambridge backers' deep interest in keeping Pocahontas pure.

Historically speaking, countless are the number of marriages between European immigrant frontiersman and native women later set aside when the man returned to the white world, and as numerous were those motivated by pure greed. Nevertheless, given Rolfe's own heart-torn laments, his addiction to the indigenous herb of understanding, and his clear devotion to his faith and the company, it seems that he married her because he believed doing so was righteous and, certainly, desirable, both personally and business-wise.

A major puzzlement of the union of Christian and Christian pagan, scientist and medicine woman, nature girl and citizen planter is how the couple navigated a marriage so mixed it almost defies thought. Which may be why little is said of it.

In a recent publication devoted to the confusions of marriage in precolonial Indian society in New England, the scholar Ann Marie Plane tells us that "in the pre-colonial world the most important and enduring links between adult men and women were those of clan affiliation and kinship, not of conjugal union."[34] Plane notes that indigenous marriages "could and did play critical social roles" in Indian alliances as in modern ones. Marriage between people of high stature "joined not just a man and a woman but entire villages and peoples, designed as they were to effect or maintain diplomatic alliances across the region." Since women had control of food production and distribution and owned the fields, houses, and most of the portable property, marriage into a wealthy and powerful clan could propel a man into prominence, leading the *weroanskaas* of his own clan to appoint him to major office.

Further, sexual openness at an early age was usual and accepted—one of the many reasons the English Christians labeled the locals "salvage"—although pregnancy either didn't occur as a result of these adventures or, if it did, somehow doesn't make it into the record. Similarly, divorce was common, and could be initiated by either spouse. Breakup of marriage did not injure children—who were clan members, not conjugal property—and while it was not undertaken as readily as dating different people, if the reasons were compelling enough, the "contract" was dissolved and new ones established. The children remained aware of their father's identity, but as the male authorities in their lives were clan men, known as uncles whether or not they were the mother's siblings, the change in the father's role from mother's lover to friend of the family was not particularly jarring.

There is scholarly debate about the primary mission of the Virginia Company: Was that endeavor dedicated to making money or to making converts? The history of the colony after the Peace of Pocahontas drew to an end with her death in England might seem to leave little doubt: it was the money and all it implies that fueled the fervor of men like Thomas Dale. It can be argued, however, that in those times separating economics from religion in European social systems was as difficult as making a similar distinction in Native societies of the time. Most likely it was a combination of both, with the merchant investors inclined toward trade and goods with which to expand their business and the clergy being mostly dedicated to amassing land, residents, and converts devoted to the Puritan vision of Anglo Christianity.

The differences in custom—sexual, spiritual, political, philosophical—and worldview were not taken into account at the time, particularly by the aliens. The evidence suggests that the Indians were more open, or at least more pragmatic, in cultural affairs, but whether this seemingly greater acceptance of difference stemmed from their relative homogeneity, compared to the religious wars still raging through Europe, is difficult to determine. The locals were also at home in the region, while the English were aliens who knew nothing of the land or the people. They were the original "strangers in a strange land."

In *The Inner Reaches of Outer Space,* Joseph Campbell offers us some insight on these matters of cultural differences that occurred then as now. He observes:

> It all comes of misreading metaphors, taking denotation for connotation, the messenger for the message; overloading the carrier, consequently, with sentimentalized significance, and throwing both life and thought thereby off balance. To which the only generally recognized correction as yet proposed has been the no less wrongheaded one of dismissing the metaphors as lies (which indeed they are when so construed), thus scrapping the whole dictionary of the language of the soul (this is metaphor) by which [hu]mankind has been elevated to interests beyond procreation, economics, and "the greatest good of the greatest number."
>
> Do I hear, coming as from somewhere that is nowhere, the frightening sound of an Olympian laugh?[35]

Which is to say that a major problem for contemporary students of colonial customs and thought is that of confusing our metaphors; interpretative malapropisms can abound if we believe that our forebears of that era, whether English or Indian, thought as we do in this. The greater probability is that their metaphor frame, or paradigm, differed sufficiently from ours that we are as likely to misinterpret as not.

We can at least allow for this and entertain the possibility that the two societies were alike in more regards than either recognized; from our point of view they were in general hard to distinguish from one another unless we suppress some evidence and overvalue other characteristics of either.

This similarity may account for how the dusky daughter of savage sires could live harmoniously with her stranger husband. Certainly he was no stranger to pagan modes of thought; as a gentleman he might well have attended the theaters in London or other urban centers. He was literate, as his writings show—letters and records kept for the Virginia Company—so it is likely that he was conversant with English poetry of the time. As Europe moved from medieval to modern, not only had its interest in scientific methods of inquiry blossomed, but its interest in curiosities from around the world, in literature and teachings from its Greco-Roman classical heritage, and in vestiges of England's pagan past was increasing dramatically as well. One of the effects of Henry VIII's rebellion against the Roman church was to open up avenues of inquiry that had been tightly sealed. Europe, particularly England, was enjoying a period of intellectual flowering that would continue for at least two decades beyond Pocahontas's death until the Puritan revolt of the 1640s, which pushed it back to an earlier, closed period of the medieval Roman church.

Along with this newfound interest in ancient worlds and modes of thought, the modern era, which had begun in Europe around 1300, had as one of its characteristics the growing influence of explicate systems. There were a number of reasons for this shift, of course. Contemporary theory leans toward the "disease model," noting that the Black Death of the mid–thirteenth century had wiped out at least one-third of Europe's population and severely shaken the survivors, traumatizing them into another mode of consciousness. The trajectory toward conceiving the world in explicate, or material, ways has increased exponentially over the centuries, accelerating in its curve with the greater distance from its point of emergence. At the time of the Elizabethan era the drift had just begun to accelerate.

Perusal of Elizabethan literature, paintings, engravings, statuary, gardens, architecture, and more mundane documents proves that Elizabethan culture did not discount the presence of "animistic" forces acting on the political world. And, given his conviction that sorcerers were bent on his destruction, James I had no doubt about the power of forces beyond the rational, or explicate, order. The idea that great storms accompanied the deaths of monarchs or social upheavals was common. The court enjoyed the services of an astrologer and looked to soothsayers and other arcane sources for advice, as did the rest of the population, according to their means. While the entire populace was, or appeared to be, firmly Christian, their Christianity featured a variety of "pagan" elements, and had so done since the coming of the Christian missionaries centuries earlier.

In Europe the plague brought on a Great Transformation; it is highly likely that the dreadful time was signaled by some astronomical event evident to astrologers and astronomers (the distinction between the two professions was not as clear-cut as it is now). Nor was the transformation confined to Europe or the Americas—where it, or *something,* had eradicated large populations and whole civilizations a century or so earlier. The Great Transformation seems to have sprung up all over the world at about the same time, give or take a century. Certainly the changes between the indigenous civilizations of the Americas in that time, continuing into the time the Virginia Company was getting established, are almost incomprehensibly vast. The "metaphors" of those lost times, which persist in Oral Traditions, petroglyphs, edifices, and archeological remains, survived in some form in the fact of Pocahontas's presence. It is likely that the Great Medicine Way, the *midéwewin,* was a vestige of that ancient era, just as Christian thought as it was transforming during the Renaissance was a vestige of ancient times in the circum-Mediterranean region.

The ensuing progress of the transformation that reached around the world was probably "written in the stars." Unfortunately, we have lost the ability to interpret astro-metaphors. Suffice it to echo Joseph Campbell: we mix metaphors at our peril; to apply explicate rules to implicate systems is as dangerous as traveling the *manito aki* when not in the proper dream state.

Meanwhile, back at Henrico, Thomas Dale and Alexander Whitaker have their convert. Pocahontas is entombed in "proper" Puritan dress, catechized, and ready for the remaking ceremony: the holy sacrament of baptism. The Reverend Whitaker gets the honor and he chooses a telling name for her: Rebecca, the scriptural mother of Muslims and Jews. "Pocahontas"/Rebecca may be grateful that it is water rather than some other element that the English use for this part of their remaking ritual. They might have required a seven-day fast and vision quest, a challenging trial often accompanied by the use of purgatives and emetics, ashes and mud ground into hair and skin, and body piercing. Smith, if you recall, had his own catechizing: an intricate affair administered by a number of *tsenacommacah* dignitaries, female and male, concluding with the ultimate ceremony where he was renamed by the Principal Dreamer of the Powhatan, Wahunsenacawh, attired in his formal regalia. Smith kept his clothing and the objects he carried—his "medicine," as his initiators seem to have seen them. Re-dressing indigenous peoples seems to be a Christian tradition of long standing, although seemingly foreign to other peoples. Before his renaming, if you recall, Smith had to undergo a frightening ritual that, to his untutored mind at least, meant his death at the hands of terrifying, towering aliens brandishing an ax and a club over his prostrate body. The quiet prayers that the reverend offered over his captive may have seemed dull when compared with the piercing trills that accompanied Smith's remaking.

Pocahontas may have awaited her initiation into English adult society with as much trepidation as did John Smith as he faced his own initiation into the *tsenacommacah,* though she may have been unnerved by anticipation rather than dread. But although she had chosen this path, doubtless at the behest of the manito and the elders, she had no more way of knowing what was likely to be entailed in being remade by aliens than Smith did. But the requirements in her case were relatively familiar; that is, they suited her worldview and her purpose, although they differed in numerous particulars from the Powhatan way. After a time of considerable education into the customs and laws of her new medicine lodge, she was required to give up her personal name, Matoaka, in exchange for her new one, Rebecca—a reasonable enough requirement. No doubt she had to listen to hours of lectures on the meaning of the name and the implied obligations; surely she understood the meaning of the name in terms of its implicate significance in the medicine or magical order of the English, which it was her intention to master. That she did not give up her powa name is worthy of note. No magician or medicine person would do so, as it was the key to entering and exiting the realm of the sacred. Without such a name a traveler into unordinary realms might never return, or might return unable to function. To her the entire process must have seemed like the much longer one she had undergone in her ascent through the various levels of accomplishment at Werowocomoco.

The new-made English medicine woman probably assumed that the men and women of authority or status around her, for all their strange hairdos and garb, had spiritual names known only to themselves, and that they had their spirit guides and a forever silent but known bond to whoever named them. It was not a subject for conversation, idle or otherwise. Modern people have certain conversational taboos; they do not speak of sexual or other personal matters openly, and many do not do so even with intimates. The manitowinini were no different: they also had conversational taboos, but theirs were spirit related rather than reproductive or economic.

The initiation of Matoaka into the English manito aki was not complete with baptism, which signifies rebirth in the spirit. John Smith had been named and given a tribe and lands—the marshy swamp of the peninsula where they had first debarked and set up quarter—as well as weroance status. Matoaka/Rebecca received a husband, a Christian surname, a plantation—gifted to her by Wahunsenacawh or the Grand Council—and a household to run. One wonders if she knew that would come next. There's every reason to suppose she would make that fairly obvious guess; it would have been the next step for a young woman in her community. She may have spent more time trying to discover the identity of her intended than the next steps on her medicine path.

Pocahontas apocrypha leads us to believe that she and John Rolfe had a thing going, and this might have been the case. Marriage for love was one of the various reasons for forming conjugal alliances among her people. However, more formal matches were determined by the sachems and elders concerned—clan matron, priestess, priest, maybe uncle. Of course, her family and spiritual guides being distant, her reproductive and matronly status would be determined by—as I'm sure she surmised—Marshal Dale and the Reverend Whitaker.

The courtship of John Rolfe is as much an enigma as is his identity. Characterized by many researchers as a pious, straitlaced Christian man, he formed an obsessive passion for the heathen Indian woman. He was horrified by his feelings, believing, with some justification, that a relationship with such as she would put his ambition, his faith, and his life in peril. James I had written with disdain of indigenous New World people, and had voiced revulsion for such liaisons, common enough among Spaniards and French voyagers and colonists.

Despite reservations, Rolfe may have viewed marriage to the converted Indian as a chance to prove his piety and further his ambitions. He had plans for a plantation that would feature tobacco rather than corn as its main product. He confided his love to a fellow colonist, Rafe Hamor, then the company's recorder, or secretary. Hamor, conspiring in Rolfe's decision to wed the Indian princess, wrote:

> Long before this time a gentleman of approved behaviour and honest cariage, maister John *Rolfe* had bin in love with *Pocahuntas* and she with him, which thing at the instant that we were in parlee with them, my selfe made known to Sir Thomas *Dale* by a letter from him, whereby he intreated his advise and furtherance in his love, if so it seemed fit to him for the good of the Plantation, and *Pocahuntas* her selfe, acquainted her brethren therewith: which resolution Sir Thomas *Dale* wel approving, was the onley cause: he was so milde amongst them, who otherwise would not have departed their river without other conditions.[36]

Hamor continues with his tale of love and glory, recounting that the journey on which he, Hamor, acting on Rolfe's request, handed

Rolfe's letter to Dale had been for the purpose of meeting with Wahunsenacawh to discuss payment of Pocahontas's ransom. The principal chief refused to meet with the party, who met with Opechancanough in his stead. The plan of the Englishmen was to return Pocahontas and retrieve the goods the English claimed for her ransom, along with the Englishmen who were hostages of the Pamunkeys'. As they attempted to parley with the Pamunkey party at Matchut, a conflict erupted. A man aboard ship was hit in the forehead by an arrow shot from the cover of foliage. Although his injuries were promptly attended to by the *chirgeon* (medic) aboard, the English retaliated. They debarked quickly and turned their fury on the village, burning about forty houses; they also killed and injured several villagers, and looted and destroyed the canoes that were on the bank nearby. The English had threatened such action if their demands were not speedily met, and they proved as good as their word.

At some point two Englishmen went ashore and spoke with Opechancanough. While they met, Pocahontas/Rebecca was able to visit with two of her brothers, who, according to Hamor, were overjoyed to see her thriving. She must have informed them of her impending, or at least, hoped-for, nuptials. Having come to some sort of puzzling kind of accommodation with the people around Wahunsenacawh, the English lifted anchor and headed back to Henrico, where they had April planting to take care of. For reasons Hamor does not explain, the English, so set to commit mayhem and major damage should their demands not be met, abruptly left the scene. But they had met their unspoken goal, it seems, as soon as Pocahontas announced to her visitors that she meant to remain among the English (and continue her mission). Their business concluded, the English and Rebecca headed back to Henrico in time for spring planting. She may have taken the time during her visit with the Indian men to pass on any intelligence she had gathered.

Soon the people at Henrico received Wahunsenacawh's agreement to Rebecca's marriage to Rolfe. Opachisco standing in for her, as was proper in the *tsenacommacah,* rather than her "father" Wahunsenacawh, which wasn't, the two were wed.

Just how Powhatan got the information [concerning the impending marriage] we do not know, nor for that matter, do we know what was to become of Dale's trading ace once his hostage was married to Rolfe. In short, the details of this important event are strangely lacking, while those provided by Hamor are chronologically adrift. . . . Hamor next tells us [after saying that the negotiations had been suspended for planting] that ten days after receiving news of the marriage, Powhatan sent an old uncle of Pocahontas's to represent him at the celebration. At the very earliest, therefore, the wedding would have taken place in mid-April.[37]

However, Hamor's record of the trip to Matchut—during which he, Hamor, gave Rolfe's letter to Dale, Rebecca spoke with her brothers, the English overreacted to an arrow hit, representatives of the company parleyed with Opechancanough, word was sent to Powhatan requesting his permission for Pocahontas to marry Rolfe, and more—was for the benefit of the main office in London. It was how the various enterprises kept the company and its stockholders informed (or misinformed). Often as not, these standard reports, similar to monthly reports prepared by branch offices for headquarters' records, were as much fabrication as fact. Equally, it was as common for reasonably factual reports to be destroyed further up the line.

However incomplete or fanciful, the records do reveal that the agendas of both parties, Dale and Pocahontas (and their respective backers), were met. Whether each was apprised of the other's agenda will never be known. However, it's not necessarily impossible for that to have been the case. When the spirit world, the implicate order, is concerned, the fixed notions of the explicit world order are irrelevant. Because there are two converging, although seemingly opposing, intentions, each backed by considerable manito power, two lines of probability emerge. One seems to suffer considerable obscurity in the events and what motivates them, while the other offers growing clarity concerning the role Pocahontas was to play in the Puritan faction's New World drama. As Rebecca she is designated the mother of their dreamed-of New Jerusalem, the "city upon a hill" defined by John Winthrop of Plymouth Colony in his sermon and speech to the Pilgrims as they prepared to disembark from the *Mayflower:*

> We shall find that the God of Israel is among us. . . . For we must consider that we shall be as a city upon a hill: . . . we are commanded this day to love the Lord our God, . . . that we may live and be multiplied, and that the Lord our God may bless us *in the land whither we go to possess it* [emphasis added].[38]

Just as the first son of Abraham, Isaac, was wedded to the Urdish woman Rebecca upon the firm guidance of God's angel, and from their union would descend the tribe of Israel, so from the union of this Rebecca and the son, at least in spirit, of the Puritan church John Rolfe would spring a New Israel, a more perfect Promised Land. Perhaps unbeknownst to the Puritans, the Algonquins too had a dream, a vision divinely inspired in which they were as invested as the Christians and new capitalists were in theirs. To them it was clearly the end of the particular cycle in which their ancient way of life had been proper. As the prophecies, the Oral Tradition, and the stars had made clear, nothing could occur in the time of Pocahontas/Rebecca's life but a world renewal. While Powhatan and Puritan would each define the New World differently, they were agreed that its time had come.

AZTEC TOBACCO
Nicotiana rustica L.
(from Kohler's Medicinal Plants, published 1887;
courtesy of the Missouri Botanic Garden)

4

Apook / The Esteemed Weed

From 1500 to 1800, Spain earned more money in the New World from tobacco than it did from gold.

—Joel Sherman

The universe is such a strange and wonderful place that reality will always outreach the wildest imagination.

—Arthur C. Clarke

Tobacco was as basic an element of religious practice as stone, fire, water, blood, and conversation with certain animals and supernatural beings, manito, who served as guides, totems, or modes of moving between here and now to elsewhere/when. Along with corn, beans, and squash, it was the most sacred of plants, and it was specifically designated as the plant of interaction between the *manito aki* and human beings, the world of the implicate order and that of the explicate, or mundane. Its use as an ally or spirit guide for connections with and within the sacred was as ubiquitous as is that of frankincense, sandalwood, and other varieties of incense used for spiritual reasons all around the world.

During the administration of Theodore Roosevelt, the ethnomusicologist Natalie Curtis collected a variant of the Tobacco Spirit (manito) myth or sacred story, which she published in her classic study, *The Indians' Book*.[1] She had to secure a dispensation from the president in order to collect the material, because it was against the law for Indian people to share their stories, songs, or any other aspect of the Oral Tradition with white people. It was also against the law for them to practice their own religions, although huge piles of their spiritual legacy were taken and cached in private and "public" collections around the world.

However, armed with presidential authorization, the intrepid ethnologist traveled around Indian Country, all over the United States, and among the works she recorded was this Abenaki—or "Wabanaki," as she inscribes it—story of First Woman.

The story takes place in the times outside of time. There was a great spirit being known as Glous'gap (Curtis renders the name "Kloskurbah"). One day (as such things are perceived in the somewhere that is not here), at the middle hour between sunrise and sunset, a young man appeared before the greatest of all *mamanantowicks,* Glous'gap. The youth announced that he had come to help the Old Manito however he could. After all, he was young and swift, so he could act as messenger and gatherer for the Old Man, as needed. He wanted to be Glous'gap's apprentice, his student, in the mysteries of Glous'gap's magic. The young stranger announced his relationship to

Glous'gap, calling him by the word that meant maternal uncle or male of his mother's clan, establishing the primary importance of Big Clan—that is, mother—relationship. By establishing the close relationship, both gave Glous'gap reason to let the young man stay and help, as well as reason to recognize their bond. The young man also said he had been made of the foam of the sea: "The wind blew, and the waves broke into foam. Then the sunshine brightened the foam, making it warm. That warmth made life, and I am that life." With these words he establishes for us his identity as a manito, a mystery being. We are now alerted, if we didn't know this before, that the story that follows has everything to do with magic—that is, with the events and forces that operate within the implicate order, the *manito aki*.[2]

The young man stays with his uncle Glous'gap, and on another midday a "maiden" appears before the two men where they are sitting. She calls both Glous'gap, the Elder being and his nephew-apprentice, as her children, saying she has come to stay with them. "And I have brought love medicine with me. I will give it to you," she says. By this she indicated that she had a ritual tradition, a woman's magical rite, and she would show them how to do it, which made her their "mother." Possession of such a rite proved her their elder in a ritual sense. She was announcing that she was not only an adept, but one whose level of ability entitled her to take up residence with the elder manito and his nephew.

So far, this account reminds us that with her appearance there are three worlds in existence: first, the world of the manito, or actually the one in which the manito order of consciousness *be's;* second, the watery world of the great seas that give birth to life in the void; and third, the plant world that quickens the meeting of fire (warmth) and water. The third being is known as First Woman, and from her body comes forth all-that-is on several of the levels of being-consciousness that surround and enfold our own. Because she refers to the representatives or personifications of the first two orders of existence as her children, we know that she preexists them and gives them life.

She tells the manito men that she has brought love medicine as her gift to them. Native people never go calling, never mind move into someone else's community, without offering gifts. Gifting is fundamental to Native life, and is institutionalized in numerous ways among all the Nations I am acquainted with. From the potlatch in the Pacific Northwest to "grab days" among the Pueblos, to honoring ceremonies of various kinds among the people of the Midwest, to a variety of thanksgiving and corn-dance gatherings along the eastern seaboard, gifting is a normal part of Indian life and it is considered a religious act. Normally one also offers gifts to plants, animals, rocks, and what all because they share what they have with us. Reciprocity among all orders of being is considered something like natural law, as great a force as gravity or magnetism.

First Woman also offers instruction to her "sons," or manito, with magical rites younger than the one she "owns"—that is, controls by right of skill and knowledge. "I will give it to you," she told them, "and if you will love me and grant my wish, all the world will love me well, even the very beasts." She does not yet tell them what her wish will be. Instead, she continues to elaborate on her gifts—that is, the gifts of the mystery cycle or ceremony she owns, which include strength and comfort to all comers.

"Strength is mine, and I give it to whosoever may get me; comfort also; for though I am young my strength shall be felt all over the earth. I was born of the beautiful plant of the earth; for the dew fell on the leaf, and the sun warmed the dew, and the warmth was life, and I am that life."

The narrative of First Woman is a sacred story—that is, it is a recipe for a ritual here. Her words are cast in the same mode as those of the young man, and we know of her origins not as some king's daughter, but as a process. Like Glous'gap's nephew, she announces her ritual tradition, the mystery tradition to which her *powa* belongs.

Eventually the young man and First Woman had children, and Glous'gap taught them—"did great works for them." When his work there was finished, the Elder moved to the North Country saying he'd stay there "until it should be time for me to come back to you." The family's numbers increased and increased until famine struck.

First Woman grew sadder and sadder, in her anguish going farther from their home territory every day, staying away for longer and longer periods.

One day her distraught husband followed her to the river's edge, where he sat and waited until she came back, many hours later. When she came, she sang as she began to ford the river, and as long as her feet were in the water she seemed glad, and the man saw something that trailed behind her right foot, like a long green blade. But when she reached the bank she brushed the blade off her foot, and sadness shadowed her again.

That evening her husband invited her to watch the sun go down with him, and as they stood together facing the west, seven of the littlest children came to where the couple was standing and, gazing earnestly up at First Woman, announced that they were very hungry. It was almost night, they pointed out, and there was still nothing to eat. Torn with grief, First Woman looked at them. Tears streamed down her cheeks as she hushed them, saying that in seven moons there would be plenty of food, and the famine would be over.

The poor man was beside himself. He took hold of her arms and asked what he could do to help her. She said, "Take my life." This was one thing he wasn't prepared to do, but she finally prevailed. After all, he and Glous'gap had agreed to grant her wish when the time came in exchange for the ritual she taught them, the *powa* she gave them. With heavy heart, he set out for the Northland to fetch Glous'gap, hoping the old sorcerer would save them from their promise. But seven mornings later he returned, grief stricken, with word that Glous'gap had admitted that First Woman was right: he must take her life. Visibly relieved, First Mother instructed him, as medicine women still do, on the sacred manner in which the rite was to be accomplished:

"After I'm dead," she instructed them, "get two men to take me by the hair and draw my body all around the field I will point out to you beforehand. When they come to the center of the field, they should bury my bones there. Then everyone should stay away from that place until seven months have passed. When it's time, let the people go again to the field and gather up everything they find there. Eat most of it. It's my flesh, and it will keep the people strong. Be sure to save some of it to put back in the ground so there will be more the next season. You can't eat my bones, of course," she continued, "but they can be burned. The smoke will bring peace to you and your descendants."

They followed her instructions, and after seven months had gone by the men went to the field and discovered it filled with tall, green plants. They tasted some of the fruit, which tasted sweet. Her husband named it *skarmunal,* corn. In the center of the corn he saw another plant. This one had broad leaves and no fruit, and it tasted bitter. This he called *apook,* tobacco.

When the crop was harvested, the amount was so large that he didn't know how to divide it. So he sent for Glous'gap, who came and, seeing the corn filling the round bins the people had built to hold it, understood who the "maiden" had been and what she had meant by her words of greeting. So he offered a prayer of thanks and then told the assembled crowds, "Now the first words the First Mother spoke to me have come to pass. She said she was born of the leaf of the beautiful plant, and that her *powa,* medicine, would be felt over the whole world. She said that everyone would love her because love was her *powa,* her medicine.

"And now that she is gone into these plants," he said, "make sure that this, the second seed of the First Mother, is always with you. It's her flesh. She also made a gift of her bones (*apook,* tobacco) to you for your good. When you burn them the smoke will make your minds fresh. Most of all, remember that these gifts came out of the goodness of a woman's heart. Honor her memory always; remember her when you eat, and whenever the smoke of her bones rises in your sight. Remember that you are all related to her and to each other, and apportion her flesh and her bones equally among you. As long as you do, the love of the First Mother will keep you strong and peaceful in your hearts and in all you do."

APPRENTICE OF TOBACCO MANITO

Just as First Woman, come from Plant Manito, and Glous'gap's acolyte, born from the sea and foam, connected to make another world and she died to renew it, so America's first lady and Plant Manito acolyte John Rolfe made a world-renewal connection in historic time. Both sets of transformational agents, mythic time and historic time, had obscure origins. It seems that this feature is significant, necessary even, in the great ceremonial of world renewal that would be channeled through them and their activities. The Oral Tradition seems to suggest, over and over, that the characters who make up our world, individuals in our understanding, are themselves what Jungians call *archetypes*. The ceremonies, for instance, re-create sacred events—literally, not symbolically. When the Sky Woman ceremony takes place (or took place in earlier times), the state of *manito aki* being was recognized in its fullness as it was the first time she fell from the sky.

More to the point in understanding Pocahontas's life, the tradition indicates that people are born who are analogs of a type. They are manito themselves in that the manito *powa* works through them and re-creates patterns, dynamics, beginning the cycle again. To illustrate with a crude example, it works like rebooting a computer, or kickstarting a motorcycle. Or perhaps the process is more akin to refilling your gas tank.

Perhaps the way this cycle or ceremonial works is that tobacco manito is a facet of Glous'gap, while Pocahontas/Rebecca (trickster and first mother of Israel) and John Rolfe (alchemist/tobacco acolyte) are Glous'gap's nephew and First Woman. By these two powerful beings connecting, the world we know came into being. Not right away, but eventually. The famine and drought (the latter was signified by First Woman's happiness when she, plant spirit, was submerged) signaled the end of one world cycle and the time when another one would emerge from the void. The emergence of this new world—or, put another, equally valid way, the renewal of the first world—was made possible by the ritual death of First Woman followed by the dragging of her body around the forest meadow, and finished in a sacred manner with the maturation of the sacred, *powa*-filled, plants. A parallel series of events is occurring in the life of Pocahontas, although her life appears to reprise or re-create in historic time at least two of the major mystical or sacred event sequences of the times before.

That said, it is not surprising, given the overall mysteriousness of Pocahontas's life, that it is as difficult to identify John Rolfe, the Englishman she married, as it is to define the "real" Pocahontas. It is commonly accepted that he was a son of the Rolfes of Heachem, East Anglia, but though accepted by default, this cannot be proven. While it is believed that Rolfe was a widower when he went off to Henrico with Sir Thomas Dale and the Reverend Alexander Whitaker, Pocahontas firmly in the trio's clutches, it cannot be proven. One of the biggest mysteries connected to Rolfe, though, is how he came into possession of the golden seeds of Spanish-grown tobacco. And therein lies a major clue to his identity.

Rolfe's role in the ages-old magical-mystery narrative, the Oral Tradition of the *manitowinini,* Pocahontas's people, hinges on his addiction and single-minded dedication to the cultivation of primal *apook,* tobacco. It was the degree of his devotion to *apook* that led Pocahontas into a formal alliance with him. Because theirs was a formally recognized alliance, and because the *apook* that Rolfe and Pocahontas planted was not the same as that cultivated by the *tsenacommacah,* their arrangement constituted a sacred rite, doubtless planned, charted, and blessed by tobacco manito herself.

Out of their alliance came the new paradigm that formed the United States and the world we know today. For upon the cultivation and distribution of tobacco rested the rise of the English to world domination; further, the tobacco produced from the union of Powhatan and the English, medicine woman and alchemist, soon became the dominant variety distributed to a global market.

So mild and tasteful was the *tsenacommacah*–Jamestown Colony blend that Spain lost its dominance of the most lucrative market the world had ever seen and soon went under, financially and politically. What better mate for a supremely adept medicine woman and agent of change than a man whose spirit was profoundly attuned to the tobacco manito during the Time of Great Change—than the alchemist/magician–tobacco acolyte par excellence? His connection to alchemical pursuits are as obscure as other details, but he pursued his tobacco dreams with scientific exactitude. The scientific method, somewhat like it is known today, was first shaped by Sir Francis Bacon, one of the members of Dr. John Dee's Enochian mystery school, who numbered alchemy among his learned accomplishments. *The Random House Dictionary* defines *alchemy,* a discipline that came into Europe from the Arab world, as "a form of chemistry and spec- ulative philosophy practiced in the Middle Ages and Renaissance and concerned principally with discovering methods of transmuting base metals into gold and with finding a universal solvent and elixir of life, . . . and, any magical power or process of transmuting a common substance, usually of little value, into a substance of great value."

The definition, particularly in its second part, aptly sums up what Rolfe accomplished. However, it is unlikely that he accomplished this magical transformation of an herb into wealth and power untold—the plantation system and the nation that grew from that base—alone. The Wizard, or tobacco manito, via the agency and ceremonial know-how of his Indian wife, made successful their joint alchemical procedure. Alchemy is also known as *opus magnum,* or the "great work." In the Hermetic occult tradition, studied and taught in En- gland and on the Continent from the Middle Ages on, it was the magical processes that yielded a wide variety of effects, from healing to world change, that the alchemists worked on.[3]

From the point of view of the manito-driven narrative that con- stitutes the Oral Tradition, the John-Rebecca marriage was a match made in *manito aki.* There are two clear facts about the otherwise somewhat obscure John Rolfe: he was addicted to tobacco, and he was in possession of a store of the finest of tobacco seeds generally owned and jealously controlled by Spain.

The religious wars between the Protestants and the Papists, the new capitalists and the aristocracy, led to a new world order: Europe and Native America, together in death and life, driven by the force of the manito narrative and the *powa* of the *manito aki,* gave birth to a new idea.

As the English saw it, every man could own land and live independent of monarchical control; for the most part, those who placed their hopes in the Virginia Company experiment saw themselves as living free of monarchical control over their religious life; rather, the congregation, guided by the local community's elders, would set the community's spiritual course. From that course other matters—economic and political, for instance—would arise. In their eyes, commoners were more capable of ruling than monarchs, and income should be generated by one's own investments, learning, and plantation management. For these men did not envision each family in its own little cottage; they were after huge tracts of land that they could farm for export crops, enabling them to operate on the growing world market. Seventeenth-century thought dictated that gentlemen—who at that time were the designated commoners—would be the landowners and investors in capitalist market economies. Laborers, indentured servants, and others—those men Smith decried as "the dregs of humanity"—were not included in the dream of ownership and freedom religion. It was not part of seventeenth-century expectations that the poor, long since dispossessed and enslaved as serfs, many of whom were just beginning to form the urban criminal classes, were part of the new social order. One of the unexpected consequences of the American experiment was the accidental enfranchisement of the working poor, beginning with that of the propertyless male.

In his *True Relation of the State of Virginia,* dated 1616, John Rolfe recorded his vision of this new Eden that he deemed assured by the peace that reigned in the region after his marriage to Rebecca, Pocahontas. "The great blessings of God have followed this peace, and it, next under him, hath bredd our plentie—everie man sitting under his fig tree in safety, gathering and reaping the fruits of their labors with much joy and comfort. . . ."

Early on John Smith was railing at the unwillingness of these "dregs," as he termed the extremely poor men who were impressed—

coerced—into English service to work. They wouldn't hunt, fish, or plant; they were reluctant to construct houses, fortifications, bridges, roads, or other parts of the structure of state. They would rather starve—or trade with the Indians—than take time out for survival activities, because they were otherwise occupied. They were digging, panning, exploring for the gold they had been assured lay in wait for eager English hands. After the first few years these same men—or those sent from England to replace the hundreds who died of starvation and illness—turned their enterprising natures to the cultivation of tobacco. Again the gentlemen Sir Thomas Gates, Sir Thomas Dale, Alexander Whitaker, Rafe Hamor, and others decried the unwillingness of such men to build and maintain the structure the gentlemen's dreams of wealth demanded. In the long run the stubborn determination of the English poor to participate in the dream of land ownership and personal wealth led beyond the American Revolution to the pluralistic democracy hammered out in the later twentieth century. At the time, their refusal to follow orders led to brutal treatment by both Smith and Dale, although contemporaneous accounts indicate that Smith was more evenhanded and less vicious than Dale.

This emerging society, one built on personal ownership of private property and a money-market economy based in the gentleman class but administered by the officers of the controlling investors and the monarchy, was the one that Pocahontas/Rebecca adopted. Via her marriage to the gentleman farmer and seventeenth-century scientist-alchemist John Rolfe, she situated herself at the fulcrum of the change of English social order. Pocahontas was an agent of change, diplomat, spy, and *powa* woman, or *weroanskaa,* and thus her position optimized the weight of her interventions in the nascent capitalist democracy. Because she was Amonute, gifted medicine woman and conduit of the manito, it also optimized the weight of manito influence on the power level of the *manito aki* within the tidewater region. In this position, Pocahontas/Rebecca's *powa* role, her hidden, sacred identity as Amonute, medicine woman of the Mattaponi tribe and Powhatan *tse-nacommacah,* gained full range and power. Because of the massive level of *powa* she exerted, a period of peace ensued. It lasted for several years, reaching beyond her death in England in 1617. It might have lasted much longer had not certain men—and, no doubt, women—of the *tsenacommacah* ferociously opposed the new state of things-as-they-are.

THE SORCERESS MARRIES THE SORCERER

Long ago in a land far, far away, a dead man counseled his daughter, Flower, to make a marriage alliance with the *mamanantowick,* shaman-ruler, of a nearby tribe. It was said that his tribe was wealthy and full of *powa,* and the dead man believed that an alliance with them would be of benefit to the people of Flower's tribe. In order to win the shaman's hand, Flower was required to prove her magical abilities in a series of dreadful tests that the sorcerer, her intended, devised. Having had her superior magical skills demonstrated to him, the sorcerer married her.

But he was afraid of her, and the council advised him to trick her so he could do away with her. Their plan involved the Tree of Light, which grew outside the Great House and was the source of the *powa* of the council and of all the *mamanantowicks* of the tribe Flower had married into. They figured that she would be drawn into a vulnerable position by her attraction to the great tree.

One day the sorcerer invited his wife out to see the tree, telling her that something wondrous was happening. She joined him and the others of the council gathered around and saw to her shock that the Tree of Light had been uprooted. Her husband beckoned her closer, bidding her look down into the chasm the uprooting had made. She did so, leaning far over the opening. Quicker than the eye could see the husband pushed Flower into the chasm, ridding himself of her forever.

With the marriage between the young alchemist-magician posing as an English commoner and soldier John Rolfe and Mattaponi Dream-Vision priestess and Beloved Woman Amonute, the relations between the boat people and the residents cooled down. The hot war that had sporadically broken out between them, caused as much by English violations of custom, persons, or property as by Native fear, suspicion, and anger at the foreigners' incursions, settled into a quiescent phase. It was the first Indian summer; peace, harmonious relations, a fair amount of cultural exchange, and increase in business arrangements for trade seemed to have entered a positive cycle. That period became known as "the Peace of Pocahontas," and as such it held an esteemed place in Virginia history until quite recently. It lasted until about five years after she died, when Opechancanough, Wahunsenacawh's brother, who had since become the *mamanantowick,* or paramount head of state, launched open war against the newcomers. This was probably the first time that open warfare rather than sporadic armed conflict raged.

The records of the period before Pocahontas was baptized, remade as Lady Rebecca, and married to John Rolfe (i.e., those that survived Virginia Company of London paper purges) offer a number of accounts of meetings between members of the two groups, *tsenacommacah* and English. Very few of these meetings were confrontational. The reports the English made to company headquarters suggest that a general acquaintance with one another deepened over those years. William Kelso reports that evidence from the archeological dig at the original James Fort indicates a number of liaisons between Englishmen and *tsenacommacah* women. These liaisons went unmentioned in formal reports, so remained unknown until the 1990s. Similarly, the many encounters between the ordinary farmers, soldiers/warriors, and workingmen and Englishwomen and a full range of Native men, women, and children were adding to the store of informal information for each group, forging a bond between them. As Lady Rebecca Rolfe (no longer Matoaka, as she had traded her adult Mattaponi name for the English name Rebecca), the wife of an English gentleman of knowledge both arcane and on the cutting edge of seventeenth-century thought, Pocahontas continued her intelligence gathering, making regular reports back to the center of the *tsenacommacah* at Orapaks and then again at Werowocomoco when the Powhatan council deemed it appropriate and harmonious for her to return.

The "Royal Court," as the English referred to Werowocomoco, had moved soon after Smith was adopted in 1609, resettling at Orapaks, up the Chickahominy River some thirty or so miles from Werowocomoco on the Pamunkey River, for several years. As Werowocomoco was just a few miles from James Fort, Wahunsenacawh and his council early reckoned that they had best put distance between themselves and the belligerent newcomers. Perhaps they knew that there was no distance great enough to secure their form of society now that the end time, so to speak, had come. But between 1609 and Matoaka's assumption of her new identity in 1614, the situation more or less settled into a seeming standoff, with occasional forays and bloody engagements across the culture line occasioned by one or the other parties crossing that line.

It might be that occasionally the *tsenacommacah* went after the English with a vengeance, then retreated for strategic reasons. Perhaps the legendary warning Pocahontas passed to John Smith concerning the Powhatan militia's imminent attack was part of Powhatan strategy, along with her various visits to the fort to bring food to the aliens during their first winter. Certainly her visits, whether to romp with her "wilde train" and James Fort boys, or to deliver baskets heaped with corn, squash, beans, and other food, along with her warning at Werowocomoco in 1609 were designed to gain her a place of trust among them, even though the Virginia Company's situation at Jamestown was unstable most of the time, and changing even more rapidly than the *tsenacommacah*'s. Pocahontas's devotion to her mission paid off, of course, albeit the payoff came several years after the initial groundwork was laid. It came with a group of Englishmen, newly arrived in Virginia and acquainted with the Indian maiden by hearsay. They believed that she was the darling daughter of the great "Impire" of the Powhatans, and so abducting her was quite a coup. Christianizing her was an even bigger victory.

With the marriage, the evident normalization of relations between the groups made Rebecca/Pocahontas's job as intelligence agent easier. No longer isolated as hostage and convert-in-training, her family members in and out of her home, she could plant and harvest information. She had a ready network of informants and couriers available. From her, high priestess of the *apook* manito, and her clan kin, Rolfe learned about tobacco planting, harvesting, curing, and sacred uses. As writer Iain Gately observes, "[T]he Indians recognized tobacco's delicacy and dedicated more effort to its cultivation than any other crop."[4] The other crops—corn, squash, and beans—provided the basis of the peoples' diet, meat and fish being added as available. The care the *manitowinini* lavished on *apook* must have been because, of the four sacred plants given to humans by First Woman, *apook,* tobacco, was a primary element in the spiritual life of the *manitowinini.* By the word of First Woman, given before she was transformed into the four sacred plants, *apook* was the peoples' means of staying in contact with her and, by extension, with the *manito aki.*

THE MYSTERY (MANITO) OF *APOOK*

Among the Algonquin, *apook* was seen as the path to the mind of First Woman—the supernatural female progenitrix of the Algonquin, according to one of their major *powa* traditions. (There are several traditions operating in tribal ceremonial traditions because each connects with a given spiritual path.) *Apook* emanated from her bones in one version of this tradition, and from her head in another. As itself, given its great power to instruct and remake humans, *apook* was itself manito, a mystery. It was (and among many traditional medicine people still is) recognized as an agent of vision, of transformation, change, healing, orientation, teleportation of objects, creation (making something from nothing), soul walking—or perhaps all of these. No ceremony was attempted without its plentiful use. No state of proper alignment of mind-body with spirit-manito was thought possible in its absence. It was as much the flesh of the ceremony, the joining of human and manito into a state of *powwaw,* shared consciousness, as incantation, movement, or any of the other liturgical elements of the formal rite.

The English reported that smoking *pissimore* was confined to men; whether or not their observation was accurate, in Europe, tobacco—at least the market, product, and methods of production—was confined to Spain. The English, while quite taken with the herb, had no seed or seed crop and little if any information about the plant in terms of business. Yet, despite the top-secret nature of both the seed and the information, Rolfe gained a goodly supply of both. Like the masterful alchemist his scientific skill testifies that he was, trained in the disciplines of chemistry and more arcane subjects, he combined the Spanish contribution to his enterprise with the *tsenacommacah's*, and planted his mysteriously gained *apook* seeds in the soil of the *manito aki,* which was both within the earth of the tidewater area and within its psyche. He learned the incantations and respectful demeanor to be used when working with *apook* from Rebecca's clansmen. He learned how to honor the *apook* manito, and his reward was a blend that within a few decades would outsell Spanish blends and fill England's coffers to overflowing with the gold the original voyagers to the tidewater region had sought.

Rolfe might have visited a few local fields and curing houses to see for himself how it was done by the Native experts, for he was an alchemist, the seventeenth-century version of scientist, inclined to experiment in an ordered, carefully noted manner with methods, blends, and, eventually, marketing approaches. He was also a nicotine addict, and the likelihood is great that he developed his singular blend and conceived of unique marketing approaches with the collusion of the esteemed devil weed herself.

That Rolfe was inordinately successful at all three is testified to by the survival of the colony, its prospering against all odds both local and those between colony and crown, and the triumph of Rolfe's own brand, which bore the first brand name in history: Orinoco. This name was more portentous than it might seem at first glance; it was, in the words of Iain Gately, "suffused with the mysteries of Eldorado as described by Sir Walter Ralegh."[5] The tobacco manito is/are not named among Rolfe's tutors or spirit guides, of course. Had Gately been one of the traditional shamans he writes about, the manito might have been the only tutors he mentioned.

As noted in Jamestown Rediscovery, volume 7, the practice of magic was taken as reality in seventeenth-century Europe. The ideas that witches could fly and that groups of pagans could levitate during sacred dances, for instance, were common themes of woodcuts of the time. The widespread persecution of people engaging in magical practices was widespread under the rule of the Spanish Inquisition and the Puritanism of James I, who as James VII of Scotland (before he ascended the English throne after the death of Elizabeth I) had thousands of people learned in the Old Knowledge executed for the practice of witchcraft. The turmoil leading up to the Enlightenment period of the eighteenth century was largely directed against practitioners of the mystical arts. It was that war on witches, to coin a phrase, that led to the development of modern sciences as we know them.

But in Rolfe's day, alchemists were as ubiquitous as pharmacists are today. (In England pharmacies are called "chemist's," as they have been known for centuries.) The queen, recently dead, had her own court astrologer, Dr. John Dee, and herself participated in activities that her nephew and heir, James of Scotland, might have had her burned for.

Looming larger in testimony to Rolfe's scientific and alchemical ability, as well as his dogged persistence, is the fact that due to his work, the plantation system gave a powerful economic base to the colonists. In a short enough time, because of a plant First Woman gave the American peoples, indigenous or immigrant, a new nation came into being. Because of the twists and turns of fortune, or of cosmic currents, that nation has become as much Indian as European, Western Hemisphere as Eastern Hemisphere.

There is the unmentioned possibility, even likelihood, that the marriage was a diplomatic and political ploy. Perhaps it was even the manito idea of a joke. John is so smitten he is in agony, by his own account. So great is his torment it seems clear to him that he is bewitched. He himself bemoans his passion, suspicious of its source. He laments his ardor, writing of his turmoil at being so enamored. He does not want to alienate his family and friends back home. He still aspires to marry a suitable woman there. Marrying a heathen, a dark savage, a devil worshiper, however willing to repent of her wicked ways, fills him with unease. He fears he will be flung into hell for his transgressions.

As we have seen, the official story nowhere suggests that Matoaka, a woman well versed in *wisoccan,* the practice of medicine, and known to moderns as Pocahontas, came willingly among the English to find herself a husband. Nowhere is the suggestion offered that she "went English" in order to ensure that a steady supply of intelligence be made available to the *tsenacommacah.* The winning fantasy scenario seems to be that she came back because she had fallen so helplessly for old blue eyes that she was forever after pining after white meat. No stories grace poem, novel, or film about how she bespelled both Smith and Rolfe and worked her *powa* on both in a sacred act of world renewal.

That she is a superbly educated ritualist—that is, magician, priest-ess—seems to have no bearing on the scholarly and popular investi-gators' deeply held conviction that white meat was so irresistible to her that the poor girl's heart didn't stand a ghost of a chance. That she was a powerful medicine woman and enchantress never seems to enter into any expert's consideration, although the popular interpre-tation misses a great deal of evidence before it.

She was in line for the leadership of the *tsenacommacah.* Given that she was intellectually gifted and possessed no small power and influ-ence among the highest levels of her own society, she was a more likely candidate for successor of Wahunsenacawh than his brother and sometimes rival, Opechancanough. Had she lived, it might have been she who became the principal *mamanantoskaa,* woman of the highest de-gree of medicine power and strategic acumen, after Wahunsenacawh's resignation.

So complete is Matoaka/Rebecca's implicate-order play, so bril-liant its execution, matched point for point in the no less astonishing maneuvers of the Indian people throughout the Western Hemisphere, that nearly four centuries later the European-based fairy tale of the doomed but civilized savage maiden is taken as historical fact. Simi-larly, it is little wonder that the *powa,* sacred, side of the story has re-mained unexplored. One is filled with wonder at the play: was it her genius alone or was it the Powers, the Mysteries, that pulled off this magical trick? Perhaps it was a combination of the two, plus a third: the alchemical and shamanist abilities of her chosen mate, John Rolfe. Ultimately, I suppose, the trajectory was inevitable, because that's what time it was. That *apook* played a major role in the match is unquestionable. Rolfe was addicted to it, experimented with it, cultivated and harvested it, cured and marketed it. His whole life revolved around *apook,* tobacco. As with Rolfe, the spiritual life of the *tsenacommacah* revolved around *apook.* While their focus was on maintaining a conscious relation with *manito aki,* and *apook* served them by making them more open to their spirit guides, cosmic currents, dangers, and out-of-the-ordinary patterns that foretold coming events, they were as devoted to *apook* as he. Pocahontas would have discovered a truly kindred spirit in Rolfe, probably without the urgings of her own guides.[6]

The part that tobacco, that most sacred of plant beings, played in the development of a colonial economy that would depend on the labor of enslaved human beings, that would make possible the American Revolution nearly two centuries later, and that would play a large part in the American Civil War is legendary. No less significant is the role of tobacco in contemporary power plays on the national scene, although the eventual harvest of that season is yet to be seen.

In the Algonquin world—indeed, in the Native world of the entire eastern seaboard as far south as Maya rule extended—tobacco played a central role in the underlying belief system that everyone, of whatever language stock and cultural difference, shared. It was if not the purveyor then the companion of boundary extension, of internationalism and multinationalism, first in this, the Western Hemisphere of the Americas, then on the east side, where it still makes its way, trickle and flood, remaking everyone in its path. Everywhere tobacco goes freedom and is sure to follow. Interestingly, health and longevity accompany the pernicious weed in her travels, all the "scientific" evidence to the contrary.[7] Or so the historical record proves. Maybe its destiny has been to remake the human psychosphere, collateral damage to individuals notwithstanding. That seems to be the manner in which it was and is used among holy people then and today. Some believe that knowledge is worth great price; others feel the same way about healing a loved one. I think that our dear Earth Woman qualifies, for some of us, as a loved one who might need a bit of shamanic healing work.

While scholars pay little attention to Rolfe's affection for his pipe, smoking and the use of snuff had become a common sight among the gentlemen at the time. Knowledge of brain science, the effect of certain molecules on brain functioning and perception, was nonexistent at the time. The market value of the "esteemed weed," however, was famous. In the light of contemporary knowledge about addiction and chemical interactions and brain activities, it becomes evident that addiction made him a perfect candidate for an infusion of love medicine—for the "sot-weed," as it was known among seventeenth-century gentlemen of fashion, was recognized as a "sacred plant" among aboriginal magical adepts and so used among them across both continents in the Western Hemisphere.[8] Rolfe's strong affinity for it indicated his equally strong connection to the *manito aki* as well as to the peoples of the four sacred plants—corn, beans, squash, and tobacco. Pocahontas, out in the wilderness of English culture on her own, would have quickly realized that and set about enchanting this man clearly chosen by the mysteries, the manito.

Sacred, a word that is much misused in our time, once had a specific meaning. That which is designated sacred is a place, activity, person, or object that is ontologically enmeshed in the implicate order. The American English term *sacred,* in the varieties of its possible locations as place or site of consciousness forms unsuspected in modern usage, is contained in Algonquin terms: manito, *manito aki, Gitchee-Manito,* the Great Mysterious Being. In that sense, *apook,* tobacco, was seen as such an object. It was one of the four sacred plants, or the four sisters: corn, beans, squash, and tobacco. In Algonquin cosmogony, *apook* enabled the user to enter myth time/space by harmonizing human brain consciousness with the "vibe" or standing wave form common to manito awareness. *Apook* made possible that step beyond the borders of the explicate into the implicate order, the *manito aki,* in which our world is enfolded like a small air pocket in chiffon cake. However, it's a good idea to remember that the kind of tobacco then in use was a far cry from the sugared, chemical-suffused stuff that is currently marketed. Additionally, this time, in a cosmic or *powa* sense, is not the same. That would mean that the dynamics of the implicate order, the *manito aki,* in itself and as it interfaces with North America are different from what they were then. The differences show in lifestyle—cities, towns, highways, population, fuel consumption, chemicals, and all that accompanies modern life—but they are more basic than contemporary analysis suggests. Or so the people of Pocahontas's time would have believed. For them, the world was not a primarily material system, composed of matter, energy, and movement. It was a manito system, composed of energy systems that necessarily included intelligence (*powa*) and information as part of the whole.

Tobacco continues to play a major role in shamanic work even today. In *The Cosmic Serpent: DNA and the Origins of Knowledge,* Jeremy Narby presents a complex and compelling argument concerning the scientific truth of shamanic research methodology. He comments, "According to my hypothesis, shamans take their consciousness down to the molecular level and gain access to biomolecular information." Narby has marshaled an impressive array of sources from shamanic practitioners and their Oral Tradition, the biomolecular sciences, and ethnography in support of his hypothesis. At one point, Narby notes that ceremonial Oral Tradition is based on the fundamental assertion that knowledge is best transmitted through images, metaphors, and stories. He might have added pictographs and other graphic arts as part of this system of knowledge storage and transmission as well. He observes that "myths are 'scientific narratives,' or stories about knowledge (the word *science* comes from the Latin *scire,* 'to know')."[9] Pointedly, Narby reminds us of "the extent of the dilemma posed by the hallucinatory knowledge of indigenous people. On the one hand, its results are empirically confirmed and used by the pharmaceutical industry; on the other hand, its origin cannot be discussed scientifically because *it contradicts the axioms of Western knowledge* [emphasis mine]."[10]

While much of Narby's study focuses on the use of the sacred plant *ayuasca,* he explores tobacco as a "teacher" or guide to scientific studies of the shamanic kind. One of his discoveries is that certain plants enable shamans to "read" or download knowledge from DNA. He is quite specifically scientific in this analysis, carefully drawing from a wide variety of sources to explore this seemingly revolutionary concept. His exploration of the deeper significance of the "flying serpent," Quetzacoatl, and the serpent imagery from points around the world, including ancient Egypt, that is virtually identical in construction as well as in significance, is a central theme in his discussion. One of his more significant realizations is that the Nahuatl word or suffix *coatl* has dual meanings: "serpent" and "twin." He goes on to examine this seeming serendipity in terms of the construction of the DNA molecule in its two-stranded, or double-snaky, form as Francis Chrichton discovered.

THE MYSTERIES OF JOHN ROLFE

One of the many meanings of *manito* is "little mystery." One of the main features of a sacred story is the mysterious nature of the major events and of at least some of the dramatis personae. Rolfe, like Pocahontas and Wahunsenacawh, three of the major players, qualifies as a mysterious character indeed. It would seem that Rolfe's widower status rides on an entry in Thomas Gates's ship record. It marks the death of a newborn girl, Bermuda Rolfe, during the time the *Sea Venture* was wrecked near Bermuda and the survivors awaited new transport to Virginia. Presumably, Mrs. Rolfe died somewhere, somehow, but establishing her death is impossible.[11] There is no mention of a Mrs. Rolfe remaining in Jamestown when John Rolfe, Thomas Dale, and Alexander Whitaker, Pocahontas in tow, headed upriver. They were feeling energized by the improved chance they had gained of increasing Puritan influence in London financial and ruling circles by netting a high-visibility convert for the cause—and, in Rolfe's mind at least, by the adrenaline rush of the prospect of developing a viable plantation system based on tobacco. Perhaps he had his own *apowa,* or Dream-Vision, about how his achievement would play in his homeland: the hero of the Virginia venture; the guy who made a sinking enterprise sail. But no records of his possible dreams have come to light, anymore than mentions of the first Mrs. Rolfe have. Maybe her death is somehow fused in his tobacco-drenched mind with the precious contraband he carried. As he secured the Bermuda tobacco seeds by stealth or some high-level duplicity, perhaps he felt that the less said about events aboard the *Sea Venture,* the better. Shakespeare might have left us some clues about these secret events in *The Tempest,* the play he wrote about the wreck of the *Sea Venture.*

Certainly the bard makes some pointed observations about court duplicity, betrayal, and the power of a great magician to get even by bidding manito, spirits, to his will.

It is not entirely out of the question to speculate about the mysterious Mrs. Rolfe, given the other mysterious elements in Rolfe's own life story. Since her name does not appear in the ship's records, although she is the wife of a gentleman attached to the company, her absence invites it. "Indian" marriages, as they would later come to be known, seem to have been rather more than less common, given the evidence of the recent excavations at Jamestown. In these, an Anglo-European man would "marry" an Indian woman, knowing the union was not necessarily recognized by the Anglo-European establishment. Thus, such a man could have a Christian or "real" wife, and an Indian woman who believed she was his wife until he left one day and never returned—or, as in Puccini's *Madama Butterfly,* returns with his Anglo-European wife in tow.

Sir Thomas Dale, a married man whose wife had remained safely in England, found himself stricken by Indian love medicine. After he had been assured that Rebecca would remain with the Puritan community as Lady Rolfe, he sent an emissary to Wahunsenacawh to ask for the hand of the chief's twelve-year old-daughter. Once again, the sturdy Hamor was drafted as emissary. Dale had married shortly before he left for Jamestown, so his married state might have slipped his mind . . . or sufficiently enflamed it, given his enforced bachelordom, that he hoped to assuage his loneliness with Rebecca's lovely sister.

Hamor makes no mention of Dale's English wife to "brother" Powhatan. Had he done so, it would hardly have mattered to a man of a people who saw polygamy—and polyandry—as reasonable marital options, practically de rigueur for men of high status and probably as necessary for woman similarly placed. It seems that matters matrimonial were, at that time, about as open to personal situations and interpretations as spelling was. Further, it seems, given both the record—or what remains of it—and the dictates of particular traditions, marital status was fluid on both sides of the pond. Not only did polygamy flourish among English and Algonquin, but polyandry, at least among the Algonquins, was likely.

Whether this practice was reserved for high-status women, women with a sacred or ritual role (i.e., priestesses such as Pocahontas), or was a common enough feature of life in the *tsenacommacah* awaits study. However, it is as not as unlikely as conventional ethnographic wisdom maintains. As my late aunt Susie Marmon, Laguna elder and counselor to Anglo-anthros, once commented to me over a delicious slice of her homemade peach pie and coffee with Pet milk and sugar, "They were more careless about those things in those days."

It is as unlikely that Pocahontas was a virgin when Argall took her to Jamestown, as that Elizabeth, the Virgin Queen, had "not known men," to put it biblically. It is quite likely that Pocahontas had borne at least one child by the Indian man Kocuom, whom some sources say she married.[12] Because of the customary path a woman took toward maturity, particularly a maturing medicine woman, it is improbable that she remained unmarried as late as age sixteen. Because children belonged to the clan rather than to an individual, and because a mother's sisters were seen as co-mothers of her offspring, Pocahontas's move to Jamestown and marriage to John Rolfe would have been a reasonably ordinary event; any child would have remained part of its birth family, with one mom absent.

The significance of Pocahontas's status as a married woman is that it was a prerequisite for her taking a full adult role in the *tsenacom-macah*. Had she still been unmarried, it seems improbable that she would be acting as corn agent for the large population at Oropaks. At sixteen, which was her approximate age when she climbed aboard Sam Argall's boat, she was practically over the hill. By then she should have been married, to someone chosen for her by her brother, clan uncle, clan matron, or whoever made those kinds of decisions in the *tsenacommacah* at that time, and she should have borne at least one child, possibly two. With these requirements for full-status member-ship, for women, met, she would have been free to pursue other av-enues, including her own path given her by the manito.

In her case it is possible that the match with an Englishman—if not Smith, then some successor—had been decided before her birth. In the ways of the Dream-Vision People at the time, it is likely that the impact she would have on the region—which would resonate for centuries, rippling far beyond the *tsenacommacah*'s bounds—was known before her birth, perhaps before her conception. Marriage and childbearing would have increased her *powa,* in their terms. In fact, it would have been essential for any major Dream-Vision undertakings.

Polyandry was not entirely out of bounds, although, like polygamy, it might have been reserved for certain members of par-ticular lodges or spiritual societies, but on the whole the old-timers had very different ideas about marriage and sexuality than monothe-ists. By the accounts of many European travelers, polyandry was widely practiced in the Americas, although given the European atti-tudes toward such social practices, it isn't mentioned as such. Instead, travelers remark often on the "licentiousness" of Native women, who seemed to have had a relaxed approach to sexual liaisons. Given Spanish, Portuguese, English, and French norms for women at the time, such freedom would be seen as prostitution, for among them women couldn't have more than one husband, although they could have more than one man. Often they would work under the as-sumption that the women were "given" to the strangers by their fa-thers or husbands. This was because women didn't have any rights over their own sexuality among the light-skinned voyagers from across the sea at that time. And like storytellers everywhere, they were far more likely to interpret events in ways and contexts familiar to them, the storytellers, than otherwise. Even today ethnologists hold that polyandry is highly uncommon; however, this perception is as likely to be because the academic mind is prepared to fix behaviors in terms it recognizes, and no one on the Native or exotic other side is likely to correct European and/or Christian-Moslem misconceptions.

WISOCCAN / LOVE MEDICINE

Another of the mysteries surrounding this period in Jamestown and the life of the Beloved Woman who would change the world in a sacred manner is her relationship to John Rolfe, and his to her. There is the suggestion that true love, shared and passionate, was not the whole of the story, if we believe the documents that Hamor and Dale filed, and read them with a bit of cynicism. The politics of this period in England were chaotic; there was much jockeying for position, not to mention survival at the hands of the king, given his penchant for committing mass murder of religious miscreants; his reputation as coarse, loud, and rude; and the nature of the times in England, when the royal palace was often adorned with heads on a pike.

It seems clear that John Rolfe is tormented by his passion for the "salvage" maiden, but the letter written in his hand, submitted to Sir Dale, and eventually included in the report to the City of London (the Virginia Company's investment group) offers a sane and sober justification for the match as a decision made for the good of the company and sound fiscal management. It is not made as clear in whatever documents remain that the match has as much to do with Puritan power plays as with mercantile ones. In either case, and the two may have been inextricably linked, the fact that both interest groups served and survived at the pleasure of the monarch played a decisive role in whatever documents survived seventeenth-century shredders.

On the other hand there is a hint in the account Smith leaves us, years after the affair and the deaths of both Rolfe and Lady Rebecca. He comments that while there was no doubt that Rolfe was inflamed by the exotic Indian princess, Rebecca's affection was mild, though her husband continued to be consumed by passion. One assumed that Smith, no eyewitness to the marriage, engaged in a bit of gossip, perhaps with some of his writer friends in London. He also seems to have had occasion to view some of the letters that Dale, Whitaker, and the enamored Rolfe wrote.

There are so many crucial pieces of the record remaining uncertain that an interpreter of this cultural legend on its way to becoming myth must proceed with care. Added to the scholarly difficulties, there is another problem even more troubling: the narration of those historical events that are known depends on degrees of clairvoyance even Edgar Cayce would find astounding.

Readers and researchers are asked to believe that we knew what a number of people, English and Algonquin, were thinking. We are asked to accept that their motives were much like ours, or at least like those put forward over the course of the twentieth century by psychoanalysis and related disciplines largely dependent on it. While it is possible that they were so, it is more probable that those people differed from us psychologically as much as they differ from us socially and environmentally, both in England and along the Chesapeake Bay. Few today would use love magic to secure national political aims, just as few royals would be initiates in soul-walking, meeting with spirits, far-visioning, or honoring earth spirits by eating, smoking, drinking, or heaping on fires a mind-altering substance believed to be divine in origin. Or so we believe.

Moreover, if we view the historical record, even in its somewhat fragmented form, through Native eyes, we see that there is sufficient evidence of a worldview operating within and upon events that is far stranger than the political, economic, and academic tides that historians take as fact.

For instance, documents show that while Rolfe laments in the throes of his war between upright purity and overwhelming lust, he records his profound fear that the devil may be luring him. Who else would bring such a pious man as he to his knees, swooning and sleepless with need, he reasons, unless, unless . . .

[Who could] provoke me to be in love with one, whose education hath byn rude, her manners barbarous, her generacon
Cursed, and soe discrepant in all nutriture from my selfe, that
often tymes with feare and tremblinge I have ended my pry-
vate Conroversie with this, Surely theise are wicked instigations hatched by him whoe seeketh and delighteth in mans
distruction. And soe with fervent prayers to be ever preserved
from such diabolicall assults I have taken some rest. . . .

Happily, the enchanted gentleman finds comfort in the idea that it
must be God, and not the "devill," because he doesn't see the love of
his life for a time and finds he still can't get her out of his mind. For
some reason, he finds reassurance in this fortunate result of his desperate experiment. He reasons that since his trial separation doesn't
deter his passion, which "in Comon reason (were it not an undoubted
worke of god)" it would, his devotion must indeed be pure, not base
carnality. Evidently, he doesn't know about "love medicine," which
was a magical procedure in common use among adepts in Indian
country of that period, as well as in those ages to come. However, one
who is aware that such magics as were in use and had great currency
among the people of this continent in the seventeenth century can
hardly avoid seeing that the evidence points to the fact that something
other than lust for the exotic and man's sublimated but turgid sexual
drives is going on. Rolfe himself, good seventeenth-century Christian
that he was—or made himself out to be—did have an inkling that
something very unchristian was going on. None of the reports from
the time remark on it, yet there was sufficient force to Rolfe's ardor
that other writers did make something of it later.

Rolfe's testimony bears witness through the centuries of the force
of his passion for Rebecca. This is not characteristic of this young
Puritan. When he writes his report for his original stint in Virginia,
he makes no reference to the death of the other Mrs. Rolfe or his infant daughter, Bermuda, although he will make note of his grief for
Rebecca and his trepidation at leaving his dear Thomas in England—
though, he explains, he felt he had no choice.

Has something happened to our obscure hero in the time between his arrival in Jamestown as a recently once-widowered man and his second arrival, widowered once again? He lost a daughter, but gained a son, various plots of land, and an economic future that seems bright indeed, it's true. But something other than the initiation of the young-man theme beloved of novelists the world around is at play in the eerie reprise. Rolfe's love affair with tobacco and his dusky princess-wife was indicative of a deeper transformation. John Rolfe was remade, just as John Smith had been, although the latter received the full treatment while the induction of John Rolfe into the ways of the sacred—English or Powhatan—went unrecorded. Rolfe's sojourn to the New World is bookended with life-changing and almost identical events.

The Powhatans had made their first attempt at remaking an Englishman when they captured John Smith. However, that initiation didn't seem to have taken, or if it did, he was dead, or so they thought. Enter John Rolfe, and, on his heels, as it were, saucy, seductive Miss Mischief, no longer adolescent but in the awesome stage of young womanhood.

Tobacco was used among them for total remaking of young men. Of course, John the Second didn't exactly qualify—but then neither had John the First—but once again the *tsenacommacah,* with its otherworldly prescience, knew a world changer when they saw one. For while John the Second wouldn't become the world-renowned adventurer-hero and thematic hero of the yet-to-be born nation that John the First would, he would become much, much more than a media icon. He would become First Father, proper counterpart to First Woman: bringer of tobacco and identity to the gestating new world.

There is no historical record of Rolfe being wined and dined, and thus purified, as he was taken all over the *tsenacommacah,* then ritually killed in order to be born again, citizen of the Powhatan Alliance. However, John Smith disapproved of tobacco use, and his aversion may explain the differences in their treatment. Then again, there's little reason for such to be recorded. The record does note that

Pocahontas's family was often at the Rolfe place, and while John the First was given the whole treatment at Werowocomoco, it was no longer their temple town. The Powhatans, ever adaptable, had moved their priest king and all the associated personnel upriver. They might as well have diversified their investment in the sacred, scattering it among various towns and centers. Given that by now Rebecca was their connection to the aliens and a priestess of no small stature, wherever she lived would have been a sacred place. I have no doubt that whatever local spiritual practices she brought to Christendom with her formed part of her new life, however pure a Christian she seemed to the hopeful Englishmen Whitaker and Marse Dale. The enchanted and distracted Rolfe was unlikely to notice much; it was all foreign to him. Powhatan religious practice differed hugely from Christian rites, of course, and the English were pretty myopic under the most obvious circumstances, even when accurate observation meant their survival.

But whether or not a formal ritual was part of the remaking of John Rolfe, his heavy indulgence in tobacco along with his intently focused work with it and his profound passion for his lovely lady wife would have yielded the desired transformational effect. I imagine the process had started sometime before he met Pocahontas; *Nicotiana tabacum*—or *apook (apooke, apooc)*, as the *tsenacommacah* called it—had gained hold of his consciousness by then, and was well on its way to remaking him.

The Powhatans, via Pocahontas, would have found a gentleman who had been chosen by the manito as well as given favor by the alien's chief, Marshal Dale, and one of their priests, young Whitaker, of great interest to their plans. They needed an ally among the strangers, whose magic, they believed, was so much greater than their own. Clearly this favorite of the gods, chosen by one of the greatest manito bestowed on Turtle Island by First Mother herself, coming as it did from her bones, would be the ally they sought. Given that he was already deep within the *manito aki* in his heavy indulgence in the sacred weed, he would have seemed a prime candidate. Given that, their agent among the aliens, Pocahontas, set about being converted to the way of greater magic, while magically seducing this paragon and bringing him to the destiny the Mysteries had chosen.

Iain Gately informs us about the tremendous sacred and transformational power of tobacco:

> Tobacco played a central role in the spiritual training of shamans. In the right doses, tobacco is a dangerously powerful drug and a fatal poison. Shamans used tobacco, often in conjunction with other narcotics, to achieve a state of near death. . . . Shamans undergoing initiation training were required to take enough tobacco to bring them to the edge of the grave.[13]

There was a practice among the Powhatans at the time the English first came to their country called *huskanaw.* It was a long, involved ceremony in which young men around thirteen to fifteen years of age were inducted into a kind of male elite. In this violent and potentially lethal form of initiation, the youngsters were given high doses of mind-altering drugs. (Helen Rountree suggests jimsonweed was included in the wicked brew as it grows wild in the region, as in many other locales where it is definitely used for this purpose.)[14] What with that, going off with the *weroance* (or "benefactor" as Carlos Castenada's Don Juan Mateus styles a man holding the office of teacher and guide to the Other Realities), speaking in tongues for days while being fed large quantities of drugs but little if any food or drink, the initiates walked very close to death indeed. Those who survived this horrific ordeal—and not all did; thus the women standing near the chosen during the ceremonial hours before the lads headed off into wilderness, mourning their sons—emerged born anew. Perhaps they thought of this birth as being born of man rather than of woman; surely someone, somewhere had constructed this second birth that way. However, their ordeal proved them eligible for positions of power and glory in their civilization as they grew in age and wisdom in the view of gods and older, more powerful men. "Only initiated men were considered sufficiently attuned to the needs of their nation, rather than of their families, to be eligible for positions as councilors to their rulers."[15]

Rountree makes another observation that bears on the idea of Rolfe as initiate of the *tsenacommacah:* "The huskanaw was a deadly serious affair for the boys and their 'keepers' alike. Absolute secrecy was essential for the regimen to work."[16] We do not know that Rolfe went a-berrying with his wife for extended periods, but as head of his household as well as a soldier in the service of Dale's militia, he must have gone abroad on occasion. Every writer on his activities notes the presence of Rebecca's family at the homestead, and many note their involvement in coaching him on the finer points of tobacco cultivation, an activity that they saw as sacred and therefore fraught with extraordinary potentials for harm as well as good. They were under no illusions about its danger; they lived in harmony with danger and with violence as they did with the seasons and whatever life brought. But being accustomed to walking a perilous road doesn't mean one has a cavalier attitude toward one's situation: sacred beings must be treated with respect and care at all times, or dire events will ensue.

One way that his seduction might have been accomplished is via an incantation that involved the use of tobacco. These charms or chants can require several days of repetition, fasting, and other methods of preparation for the ceremony, and the agreement of certain spirit people to give the officiant aid. The Cherokee, some of whose towns were near the tidewater area, recorded some of their incantations. A sampling of incantations was collected and recorded by a Cherokee couple, the linguists Jack Frederick Kilpatrick and Anna Gritts Kilpatrick, in *Walk in Your Soul: Love Incantations of the Oklahoma Cherokee.*

One goes as follows:

The White Tobacco has just descended to you and me from Above.
The Wizard fails in nothing!
The White Pipe has just descended to you and me from Above.
The Wizard fails in nothing: you and I are never to part!

The Kilpatricks go on to explain:

> One cannot be absolutely certain who the "Wizard" is that "remakes" the tobacco which, when smoked in the "White Pipe," will bring lifelong conjugal contentment. Spirits ["The Wizard who fails in nothing"] are no less willing and able to perform if not addressed by name. . . . English translations give few hints of the frustrating difficulties in both memorization and pronunciation that some of the medicine man's professional *idi-gawe:sdi* present. They abound in tricky quasi-repetitions, and the language used is sometimes as far removed from contemporary Cherokee dialects as is Chaucer from Joyce. To "remake tobacco for women" is the caption upon this Gha:hl(i) Tso:dalv incantation.[17]

Of course, this example is from the Cherokee, not the Powhatan, and it is from the Cherokee in exile (they were removed from their homelands in the American Southeast in 1824 by force at the order of Andrew Jackson, then president, by the local militias). However, the Cherokee were neighbors of Pocahontas's people during her lifetime, and the incantations in use by Oklahoma Cherokee practitioners of traditional rites are, as the Kilpatricks note, ancient in origin and in present form. While the particular rite she used to assure conjugal contentment could not have been the Gha:hl(I) Tso:dalv incantation (that was for seducing a woman, for one thing), whatever she did must have resembled the brief example just given.

THE ALCHEMY OF SMOKE INTO NATION

It has been remarked, with more than a bit of irony, that "Virginia became a relatively permanent, stable plantation with a secure future only when it began to build upon smoke."[18] While the authors of this remark don't notice some of its implicate-order resonances, they are there. The Indians of the Americas invariably accompany ceremonial actions with smoke. Whether they smoke a pipe, throw tobacco on a fire, light sage, sweetgrass, copal, cedar, or some other consciousness-raising item, the ceremony cannot be done without it. The same can be said, of course, of most sacred actions across the world: joss sticks (China), frankincense (Western and Eastern Europe), sandalwood (India), and so on. Making sacred (which is the literal meaning of "sacrifice") is always accompanied with smoke. The spirits—whether manito, Holy Spirit, dginn ("genie"), daemon, sprite, angel, or other species of the world beyond and throughout this one—can communicate only when certain conditions obtain. These conditions are largely to do with the mental state of the participants and officiants in the event, although having an area that has also been "smoked," and thus "raised" or "blessed," is often seen as essential.

When the Virginia Company's Jamestown residents turned their efforts to establishing themselves permanently on this soil for the cultivation of tobacco, depending on smoke to root them firmly in place, their chronic decline ended and their hold grew secure.

Virginia as a new enterprise of an as yet to be grown nation does not begin to show viability until First Mother's spiritual presence became established there. Corn, her flesh, is fundamental to their physical survival; of that there is no question. They have developed sufficient connections with the surrounding Native communities to, one way or another, be well supplied with it. The London company frequently reiterates its demand that the Virginian planters concentrate on developing a food base that is independent of the Native population, but their demands are ignored. For the most part the founders of the United States chose to concentrate their efforts on what James I and his faction judged to be the demon weed.

James I was no friend to Sir Walter Ralegh, which might explain his vehement opposition to tobacco. James I was a staunch purist of Protestant belief. Any truck with matters smacking of "deviltry" aroused his pious fury; it was for purist—that is, Puritanical—reasons that he ordered the mass slaughter of demon worshipers, as he saw it, in Scotland. His situation was somewhat different in England, for he succeeded Elizabeth I, and she had been involved in circles that concerned themselves with the very practices that James I despised. But in the early years of his reign in England he seems to have judged these practices harshly, but seldom acted too violently on miscreants, unlike his policy of extermination earlier in Scotland. He consigned Walter Ralegh and the Earl of Northumberland, the renowned Wizard of Northumberland, to the Tower.

Ralegh was one of the earliest purveyors and devotees of tobacco; like the Earl of Northumberland, he was also deeply involved in the Enochian magic circles headed by John Dee and Elizabeth I. William Shakespeare and Francis Bacon were also members of that signal cabal, and the raging persecution of dramatists, theaters, and scientists that marked the Puritan reign of terror under Oliver Cromwell indicates that their connections to a level of reality that the new religion greatly feared was the motivation for the maniacal destruction and persecution visited upon them. It might be worth noting, as well, that theatergoing, the use of tobacco, and other "decadent pursuits" were much affected by the London "dandies," as gay men of the time were known. They were, and still are, also known as "faggots"—those pieces of fuel used for a living witch's pyre.

Enochian magic, still practiced in Europe and America, is also known as angelic magic. Based in the magical traditions descended from the biblical great wizard Enoch, it instills in its devotees the belief that the incantations and ritual actions prescribed in the tradition evoke otherworldly denizens. They hold that the beings they contact, whose presence they ritually evoke, are angels, and that these most powerful of supernatural beings both shape and influence events in the material world. It was this idea that Shakespeare explores—and takes for granted—in *The Tempest,* where he clearly connects Enochian magic with settlement of the "brave new world, that has such beings in it."

It seems that King James I, the man who authorized the first English version of the Bible, was a militant Christian who had involved himself vigorously in the Scottish church's battle against the Antichrist. Before he ascended the English throne in 1603, he became King James VII of Scotland, when his mother, Mary, Queen of Scots, was beheaded at the command of her cousin Elizabeth. Elizabeth had abducted James in his early childhood and raised him as a Protestant, his mother, Mary, being Catholic. James, heir of both women and the first radical militant Protestant head of state, was convinced that Satan, or the Antichrist, or Satan through the Antichrist, was determined to take his throne from him.

After all, something of the sort had happened to his unlamented mother, Mary, Queen of Scots, although in her case Satan worked through the authority of Elizabeth I. Unlike his more clearly Puritan compatriots, however, James I was a fierce advocate for the right of kings to control the state church; perhaps he believed that only in that way could he secure his safety from the fearful Satan. Given the fact that the English Puritans, under Oliver Cromwell, beheaded James I's successor, Charles I, perhaps James's paranoia was grounded in political realism.

James took personal interest in the discovery of witches, overseeing their torture and editing their confessions. His victims included a Dr. Cunningham, implicated in a demonic plot to drown the king while he was on a sea voyage, whose fingernails were pulled out, eyeballs impaled with red-hot needles, and who suffered the infamous torture of the "boots," in which his legs were encased and beaten so that "the blood and marrow spouted forth in great abundance, whereby they were made unserviceable for ever." An average of four hundred witches per annum were burned in the latter years of James's Scottish reign. James wasn't a battlefield king, as had been so many of his predecessors. He preferred instead "spiritual wars." During the later years of his reign in Scotland he ordered the burning of an average of four hundred people each year.[19]

One of his first royal acts as James I was to outlaw witchcraft in England. The Satan-plagued king had reasons beyond the political for his rage against tobacco. The plant had a tie to a number of plants used by voyagers into alternate realities since time immemorial. Two of them, at least, were used by local English witches, or so local Christian lore held. One was henbane and the other belladonna, both of the nightshade family, as tobacco also seems to be. "Witches"—that is, Wiccan ritualists—would make a paste of one or the other and with it anoint their bodies, giving them aerodynamic potential.[20] This seemed to work, as most English believed that witches could rise into the air and travel. The story of the magically powerful English nanny Mary Poppins, in its original more noticeably than in its Disney version, refers to this widely held belief, as do the Harry Potter novels. Both have achieved great fame, as has *The Wizard of Oz*. While the readership might seem to be confined to children, it is not. The number of film and stage treatments of *The Wizard of Oz*, the continuing popularity of Mary Poppins–type tales, and the readership and film audience for Harry Potter movies testify that the idea of levitation and extraordinary modes of consciousness-being is one that doesn't easily give way to logical-positivist, mechanical-materialism versions of reality. Indeed, it is exactly these works, in print as well as cinema, that make possible a coherent understanding of the world that Pocahontas lived in and took for granted.

MANITO AKI, OZ, AND THE LAND OF FAERIE

In each of the works just mentioned, there is a space, realm, or land where magical events transpire. Whether these realms remain hidden from "rational" materialists by choice, as in Harry Potter, or because that's how it is, as in Mary Poppins, or because they must be entered via Dream-Vision, as in *The Wizard of Oz*, the idea is firmly embedded in English, Continental, American, and American Indian consciousness. The world that Pocahontas inhabited, like that of the common English-European of the seventeenth century and earlier centuries, took for granted the existence of such spaces. Both the English spiritual voyager and the Algonquin one could enter sacred space and meet with those who live there most or all of the time. Indeed, there is not a Native tradition in which such goings-on are not taken as given. The idea that such encounters, such crossings, and such spaces are false, childish, wishful, or based in fantasy is anomalous in time and across the globe. So anomalous is it that it—the idea that the universe is confined to the material plus something called energy that moves that material around—could be seen as abnormal.

A major force in compelling the turn away from the arcane, mystical, and other-realm-oriented attention of common folk and emerging bourgeoisie, James I saw a deep connection between the esteemed weed and witchcraft. "In addition to its visible hell-fire associations when smoked, the tobacco plant had a family tie to witchcraft. Witches' favorite potions for flying were compounded from tobacco's cousins . . . belladonna and henbane."[21]

For a variety of reasons having as much to do with the times as with his religious extremism, James looked upon tobacco as an ally of the great evil he feared, sorcery. James, a devout Christian although not, strictly speaking, a Puritan, did not confine himself to railing against tobacco's use among the courtiers. He penned a furious condemnation of the habit, a condemnation he fully expected all men of Christian virtue to heed. James's *counterblaste,* vituperative from first to last—he had some skilled rhetoricians on his staff—had some equally enlightening observations to make about Indians.

And now good Countrey men let us (I pray you) consider, what honour or policie can moove us to imitate the barbarous and beastly maners of the wilde, godlesse, and slavish Indians, especially in so vile and stinking a custome? Shall wee that disdaine to imitate the maners of our neighbour France (having the stile of the first Christian Kingdom) and that cannot endure the spirit of the Spaniards (their King being now comparable in largenes of Dominions, to the great Emperor of Turkie) Shall wee, I say, that have bene so long civill and wealthy in Peace, famous and invincible in Warre, fortunate in both, we that have bene ever able to aide any of our neighbours (but never deafed any of their eares with any of our supplications for assistance) shall we, I say, without blushing, abase our selves so farre, as to imitate these beastly Indians, slaves to the Spaniards, refuse to the world, and as yet aliens from the holy Covenant of God? Why doe we not as well imitate them in walking naked as they doe? in preferring glasses, feathers, and such toyes, to golde and precious stones, as they do? yea why do we not denie God and adore the Devill, as they doe?

Now to the corrupted basenesse of the first use of this Tobacco, doeth very well agree the foolish and groundlesse first entry thereof into this Kingdome. It is not so long since the first entry of this abuse amongst us here, as this present age cannot yet very well remember, both the first Author, and the forme of the first introduction of it amongst us. It was neither brought in by King, great Conquerour, nor learned Doctor of Phisicke.

With the report of a great discovery for a Conquest, some two or three Savage men, were brought in, together with this Savage custome. But the pitie is, the poore wilde barbarous men died, but that vile barbarous custome is yet alive, yea in fresh vigor: so as it seemes a miracle to me, how a custome springing from so vile a ground, and brought in by a father so generally hated, should be welcomed upon so slender a warrant.[22]

Shortly before James inherited the English throne, the famous poet and dramatist Ben Jonson wrote a satire that addressed the issue of adulteration of the tobacco Englishmen were buying from Spain. His play, a masterpiece titled *The Alchemist,* seems to have provided focus for Puritan leaders' distaste for the weed. In a 1602 antismoking tract that Gately rightly judges "doggerel," *Worke for Chimney Sweepers,* the most impoverished of the Queen's subjects were advised:

But hence thou Pagan Idol; tawny weed.
Come not within our Fairy Coasts to feed,
Our wit-torn gallants, with the scent of thee,
Sent for the devil and his Company,
Go charm the priest and Indian cannibals,
That ceremoniously dead sleeping falls
Flat on the ground by virtue of thy scent.[23]

Heavy language indeed. It needs to be pointed out that one man's devil is another's manito, and much of the slaughter of the innocents around the world was occasioned by exactly that great difference in interpretation of the same experiences and facts. It must be added, in defense of the Europeans, that the centuries-long dreadful encounters with the Black Death, the most recent of which had decimated Europe, had made them fearful of forces they couldn't control, couldn't see, and couldn't explain in terms other than religious ones.

The fourteenth-century wave of plague killed millions; to get a grasp on the extent of the horror, imagine a plague now that wiped out 125 million to 140 million people—in the United States alone. Their deep fear of the demonic was perfectly reasonable, given the state of their knowledge, given that the Great Plague killed huge numbers of their most clear-minded citizens. The ensuing consequences, which included burned towns and walled-off portions of cities, left centers of learning and ordinary interchange abandoned as people died or fled the cities and towns; all had lasting effects. And those effects directly affected Indian America, north and south, as they affected the Pacific Islands and, in different ways, Africa.

It is no accident that the great rising against the Roman church and the Papacy, the rise of mercantilism, and the waning of feudalism occurred in the aftermath. World change is never a pretty thing. With it come upheavals we safe modern readers and writers can't foresee. Also no accident is the fact that by and large the decimation of the people of the Americas was a consequence of disease. The patterns of chaos are as precise and orderly as those of the state of lesser dynamism known as "order." The elders in Pocahontas's time would know that chaos and order both have their time and their place, and when it's time for the one, nothing and no one can stay it. It is likely that Pocahontas, an initiate of the Great Medicine Lodge, had been instructed in the great cycles. It is equally likely that, a woman of power in her own right, Matoaka had seen enough in the *manito aki* to realize the truth of what she was taught.

· · ·

Rebecca, who in her identity as Pocahontas acted as a corn agent for the *tsenacommacah,* becomes a tobacco agent, a "sot-weed factor," for the *manito aki.* Her history, when decoded in terms of her life as representative of the sacred tradition, documents her identity in the great ceremony of change and renewal: she is First Mother. She is corn woman and tobacco woman; not as grower, but as agency by which the tradition, which is not a possession of the *tsenacommacah* but of its forever self, is sustained into our time. In other words, Rebecca, as her Christian name acknowledges, is "the mother of us all," and that identification, made centuries after Jamestown and the establishment of Virginia Plantation, solidifies the matter, making it explicate. Also, in other words, the *manito aki* isn't a cultural thing, as in "it's my culture," any more than the land is the belonging of some landholder, as in "it's my land." No, it's not. The earth is. Period. Just as the *manito aki* is. Period. No one owns it, no one controls it, and no one tells it what to do. The other world does as it does, just as the earth does, although it is the pose of modern humans to imagine that we can control either and both, that both are but an extension of our egos.

The history of the *tsenacommacah* in Rebecca's time bears testament to that. They knew that it was time, that in the implicate order the inner worlds were changing. They knew the significance of that fact: their world was—like it or not, ready or not—changing. They knew that they as they were then constituted would not survive, because they knew that their existence was an expression of the Greater Order of all that is, and that when that order recalibrated, down fell the *tsenacommacah,* cradle board and all. This they knew, and this Pocahontas/Rebecca knew also. How could she not? She had been given, by the inner order, a particular configuration of consciousness and physiology. She was attractive, clever, highly intelligent, and from before birth possessed of a high-order spiritual identity. It was that identity that made her the darling of the Powhatan's heart, not her role as his darling, bouncing baby girl.

Old Wahunsenacawh was far too advanced an adept himself for pure daddy delight to determine his actions. While he might have loved his kids as profoundly as most old-time Indian fathers I've known do, still of the first importance was the business of the manito order, which was a business he excelled at. We can be certain of his mastery of the *midéwewin* because he bore the title Powhatan, Chief Dreamer of the *tsenacommacah,* the Dream-Vision People. Given that, he may indeed have been Matoaka's father—her spiritual father, her mentor and master-teacher in the world of the manito.

As the first part of this study of Pocahontas, Beloved Woman of the Dream-Vision People, is connected to a major creation myth of her people, this part is connected to the story, in whatever form it took around the *tsenacommacah,* of the First Mother, she who gave both corn and tobacco to her children so they could share and thrive. Central to her story and the ceremonials that had to surround and inform it is the power of love.

Now, it is all too easy for moderns to think about love songs and peace songs and the Beatles and ten-hankie movies when the phrase "power of love" is mentioned. But that is not what is meant by its use (in the translation of the old story, mind) in a mythic narrative. Myth by its nature is sacred in and of itself. That is, when intoned in its proper rhetorical and linguistic form, it is powerful. It imbues gross matter with energies or *powa* well beyond what the material object might otherwise possess.

Roman Catholics will understand the idea here. A priest, empowered by the sacred power transmitted to him at ordination by a properly ordained bishop, bears in his hands, when they are used in a particular series of gestures and accompanied by particular words, the power to transform flour and water mixed and pressed into wafers into something that is, in essence, the same as the flesh of Christ. He can in the same way transform ordinary table wine into the blood of Christ.

One of the major differences between Protestants and Catholics—or Papists, as the mainstream Englishmen were wont to call them—was the belief of Catholics that a mundane object can, under certain conditions and through the power an initiate in certain mysteries possesses, be transformed from one state or condition of being into quite another order. Nevertheless, that is what Catholics mean by "sacrament," and in the Catholic context the word *sacred* means pretty much the same thing that it means to true pagans.

The disparity between the Roman and the Puritan churches had not become as great as it would. The institution that, after James, eventually became the Church of England remained closer to Roman Catholicism than varieties of Protestant Christianity would in time. At the time of Rebecca Rolfe, however, both the English and the *tsenacommacah* understood spiritual and sacred matters in much the same way, although how they judged them was, as all the records testify, very different. You might say that one world's god was another's devil as one world's proper sexual conduct is another's repression.

RISE AND FALL—CYCLES OF CIVILIZATIONS

The events of the next few centuries suggest that the ways of the pagan Native people were overmastered by the emergent secularism heralded by the coming of the Protestant revolt. Perhaps the old ways faded with the old forests, but they made their indelible mark on the embryonic new republic. In the world of the implicate order, the clear fact that the irrepressible Gately sums up is: "Love and tobacco succeeded in accomplishing what sermons and orders had failed to achieve; a self-supporting English colony in America."[24] Love medicine and science germinated something tyranny and zealotry could not: the embryo of "a new nation, conceived in liberty, and dedicated to the proposition that all men are created equal." More, when magic and science married, a new nation was conceived, one that slowly, as the Great Ceremony unfolds, finds itself dedicated to the proposition that all beings are created equally valid, equally valuable to the prosperity of the whole. And although the long period of gestation, and the bloody throes of birth, maturation, and adolescence, have been tumultuous and often terrible, still this great tree of liberty, of freedom, grows. As the wise people know, magic always has its price, and the greater the magical transformation, the ritual, the more it costs. This is not a religious or arcane notion, but a simple fact: energy expenditure depends on fuel, and the fuel used must be replaced for the being, whatever it is—star or human—to continue to burn.

We know that tobacco is sacred because it is dangerous, even lethal. Equally suggestive of its nature is the conflict that surrounds it over the ages. In the sixteenth and seventeenth centuries the forces of piety and health were ranged against—what? Nascent merchandising techniques? Perhaps it was none of these. The force of tobacco that stood and still stands opposite the piety-and-health lobby—and the two are as much one now as they were then—is the transcendent nature of the plant. It is not an accident, nor is it even a primitive misunderstanding of pharmacology, that caused Native people of widely varying local languages and cultures to attribute its existence to a goddess.

One account that Frances Mossiker records, also from the mid-Atlantic, tells of a party passing through a remote valley. They encountered a manito being, larger than life, sleeping there. She awoke as they watched, and indicated to them that they should return to this place a few months hence. Returning to their homes, they reported their encounter of a sacred kind, and of course the council and the spiritual leaders instructed them to follow her instructions, which they did when the time came. There in the valley they found three kinds of plants growing: corn, beans, and tobacco. The corn stood where one of her hands had touched the earth, the beans where the other had, and tobacco where her legs met her torso, a most sacred place indeed, and one that a person nears at their peril. And so it is: corn, beans, and squash sustain our bodies, while tobacco offers our spirits entry to a most dangerous, life- and passion-dense, being-place.

Of importance, it seems, is that however the story of *apook,* tobacco, is told, it is always seen as feminine. In the Abenaki story that appeared earlier, it was also associated with bonding, relationship, nurturing, and egalitarianism—all of which seem to inevitably accompany feminine-consciousness-based social systems. Certainly the Powhatan, the Dream-Vision *tsenacommacah,* was such a society, although the *powa* outlines have been obscured by English records. These records, of course, reflect an English point of view. And not only was it English, and thus unaware of Powhatan life and thought in any depth; the reporters' point of view was necessarily part and parcel

of the masculinist system it reflected. The feminine-based consciousness of which I speak is fairly well encompassed in the narrative about
First Woman.

The egalitarianism factor is contained in her instructions that the
produce of her body's remains be apportioned equally among the
people, her children. Implicit in her instruction was that this mother
did not view any one child as more valuable or more entitled to food
or contact with the *manito aki* than any other. Each person was of
equal value to every other; female and male, child and adult—whatever their status, access to basic necessities was to be universal. Among
the Algonquins this rule was enforced and maintained by the structure
of their society, in which descent and family membership depended
on one's mother's clan, and the produce of the fields and the produce
of the hunt or fishing expedition were distributed by the women.

Further, the underlying policy decisions were made by the medicine
women's council of elders, forwarded to the men's Great Council, and
there implemented. Somehow these "prescientific" people knew that
female brain function is holistic and balanced. (In fact, the electrical
impulses scientists recorded jump from one hemisphere to another in
a pattern that most resembles Christmas lights that blink on and off
randomly.)[25]

However improbable it may seem to modern Western thought at
the present stage of knowledge about Native sciences, it was this
feminine-pattern reality that most of their systems were founded on.
Those Native social systems that didn't in some fashion regard female
modes of perception, expression, and social interaction as the foundation of their society, such as the Aztecs, fell quickly into a pattern of
dominance and submission that led to their downfall at the hands of a
woman of power, Malinalli, *La Malinche.*

John Rolfe, the alchemist-scientist, seems to have known, or at
least sensed, the sacred nature of the esteemed weed. He studied the
plant in its varieties both as a plant and as to its cultivation. He did his
research among the local Indians and with regard to Spanish practices.
He had managed, quite mysteriously, as it happened, to secure a
rather large sample of "golden seed" from the Bahamas—a contraband

that would have merited the instant sinking of any English ship had the Spanish known it was being carried aboard. In fact, maybe it was possession of this precious commodity that led to the *Sea Venture*'s disaster at sea, and the death of his wife and daughter there. After all, use of the sacred always carries a price.

While there is no more reason to suppose Rolfe was an alchemist than there is to suppose he was one of the Heacham Rolfes or had attended either Cambridge or Oxford—that is, no reason at all—it is undeniable that he was more than literate, and that he was a trained scientist. At that time the two strongly suggest the possibility that he was a trained alchemist, because science at the time was a part of alchemy, as mathematics was part of astrology. (*Alchemy* is a form of the same stem that *chemistry* and *chemical* come from. Astrology and astronomy are similarly related in origin as well as focus.)

The Columbia Encyclopedia (5th ed.) defines *alchemy* as an "ancient art of obscure origin that sought to transform base metals (e.g., lead) into silver and gold; forerunner of the science of chemistry." You see, alchemy as the practice of transformation is a practical activity: by it art becomes science! The entry goes on to say that "the histories of alchemy and chemistry are closely linked" and then points out that in its guise as chemistry the quest for a method to transform base metals into refined ones is finally achieved. Or as a contemporary of the father of modern science, Lord Francis Bacon, wrote, "What's in a name? A rose by any other would smell as sweet." Alchemy, chemistry, science, magic—the difference between them may be more in our assessment of them than in anything intrinsic to the pursuits.

I have it on good authority, though, that magic is evil. The man who so informed me—quite solemnly, I might add—was objecting to my use of the term *magic* when referring to Native practices that fall within their "sacred" or ritual traditions. He, the well-known Swedish scholar Åke Hultkrantz, and the famed Mercia Eliade had all agreed that *magic* was a term that pointed to satanic or evil practices, while *medicine* or *sacred* meant activities devoted to healing.

My objector's authority for his position was a bit questionable, and seems to have its historical antecedents in the struggle for dominance on the social and political scene that was being waged between the Protestants and Catholics on the one hand, and the Puritans and monarchy-oriented Protestants on the other. During the seventeenth century and culminating in the eighteenth, the politically charged conflict over good and evil intensified. The Catholic belief that paganism and heathenism were, at base, diabolical grew in virulence. During the sixteenth and seventeenth centuries, the centuries in which the Inquisition was most publicly active, Protestantism was on an ascending curve that would sweep it into political and economic dominance in the nineteenth and twentieth centuries. It was the dichotomy articulated by the good professor and, according to him, subscribed to by other major scholars in religious studies that led to the slaughter of Native people at the hands of Catholics and Protestants alike. Fear of the satanic seems to grip the Western world from time to time. When it rises into dominance, history indicates, mass murder and genocide are the result.

How and where Rolfe learned either of those skills—both rather rare for the time—is unknown. It seems that before he appeared at Jamestown with Argall's passengers and crew, in the military employ of Sir Thomas Dale, he didn't exist. According to Mossiker, the attendance of John Rolfe at some likely institution of learning has yet to be found, as is proof that he is of the Heacham Rolfes. Perhaps he is a kind of reverse Jacob Boheme (1575–1624), one of the better-known alchemists of the seventeenth century. Boheme, according to one of his assistants, suddenly disappeared into thin air one day, while he was engaged in a particularly potent alchemical rite. He was there, then he wasn't, according to sworn testimony of his apprentice. Nor was he ever seen again. Perhaps Rolfe did the opposite: Poof! There he was! No mother, no father, no formal education, no records to back up his existence prior to the *Sea Venture's* tragic journey. A vanished wife, a recorded daughter with the very unchristian name of Bermuda—which was the name of the precious tobacco seeds he carried—and a profound devotion to a New World sacred plant.[26]

Of course, we know he did learn these disciplines, as his outra-geously successful experiments with tobacco attest. We know about his instrumental work in tobacco cultivation, and we know that someone named John Rolfe married Lady Rebecca, née "Pocahontas"/Matoaka, either at Henrico or Jamestown church. We know that he existed, because there are letters and official reports bearing his signa-ture, and reference is made to him in many of the reports and mem-oirs and letters from that time. There are pictures of him, woodcuts, executed while he was young—perhaps when, with his wife, Lady Rebecca, he visited England. During their stay they visited the an-cient estate of the Rolfe family, at Heacham, East Anglia. A plaque dedicated to "Pocahontas, Indian Princess," commemorates her visit there. Yet his origins remain obscure. Maybe he was a manito, come to implement the procedures deemed appropriate by the manito grandmothers during that great period of world change.

Despite the obscurities of his background, Rolfe's place in history seems secure: he is known not so much as the husband of Powhatan's darling daughter, but as "the father of tobacco," a title that is about as hard to grasp or justify as Columbus's honor as having "discovered America," even though it makes a certain kind of sense. Given his alchemical-scientific abilities, perhaps he did a profound ritual with the seeds before he planted them, while they were gestating, and dur-ing the harvesting and curing stages of production. Perhaps his work was combined with some incantation rites known to Pocahontas and her clan or colleagues, and together they literally "remade" the native plants, Caribbean and *tsenacommacah,* into the blend that took over the world.

All of which suggests that neither history nor legend is stranger to the power that myth—that is, narratives of *powa* power—exerts over human thought. If one looks long and hard, one can discern the working of that force and see the bare outlines of the entity we call "myth and ceremony"—as though they were words and gestures and no more.

Rolfe might have become acquainted with an early report about tobacco and its use at the earliest English attempt at making an English settlement-cum-enterprise in that part of the Atlantic seaboard now designated as Virginia, which was what they intended as the name of the entire holding they envisioned. They thought that Virginia, a great colony, would be some four hundred miles in depth and would run from the Atlantic to the Pacific. Of course, they had little idea of the extent of the actual distance between the oceans bordering North America. That colony, named Roanoke, as the more northerly one was called James Fort and later James Towne, had been dreamed up by Sir Walter Ralegh. Like Columbus, he convinced a queen to fund the enterprise, and thus the first real venture of England in the New World began. The colony was doomed, all of its settlers lost somewhere in the vast forests or waterways of the area. The men who came to the *tsenacommacah* inquired about possible knowledge about the survivors but seem to have drawn a blank. However, records of the sixteenth-century phase of the venture devoted some space to the subject of tobacco.

There is an herb called uppowoc, which sows itself. In the West Indies it has several names, according to the different places where it grows and is used, but the Spaniards generally call it tobacco. Its leaves are dried, made into powder, and then smoked by being sucked through clay pipes into the stomach and head. The fumes purge superfluous phlegm and gross humors from the body by opening all the pores and passages. Thus its use not only preserves the body, but if there are any obstructions it breaks them up. By this means the natives keep in excellent health, without many of the grievous diseases which often afflict us in England.

This uppowoc is so highly valued by them that they think their gods are marvellously delighted therewith Whereupon they sometimes make hallowed fires & cast some of the powder therein for a sacrifice: if there is a storm on the waters, to pacifie their gods. they cast some up into the air and into the water . . . also, when they set up a new weir for fish, they pour uppowoc into it. And if they escape from danger, they cast some into the air likewise: but all done with strange gestures, stamping, sometimes dancing, clapping of hands, holding up of hands and staring up into the heavens, uttering therewithall and chattering strange words and noises

While we were there we used to suck in the smoke as they did, and now that we are back in England we still do so. We have found many rare and wonderful proof of the uppowoc's virtues, which would themselves require a volume to relate. There is sufficient evidence in the fact that it is used by so many men and women of great calling, as well as by some learned physicians.[27]

Iain Gately mentions a number of the ways the herb was ingested: while smoking via pipe or cigar was the commonest method, users also rolled leaves into moist pellets or small packets and attached them to their heads just behind the ears. Ingestion of tobacco chunks by tucking them between upper or lower lip and gum, or tucked into the cheek. Brewing tea—which allowed the addition of other consciousness-altering substances and sacred additives such as fluids from a dead shaman's body—made it the method of choice for most practitioners of the sacred.

Gately comments: "A tobacco shaman used the weed in almost every aspect of his art."[28] Many shamans use tobacco smoke for diagnosis of illness, and some for treatment. Some use the smoke for "scrying"—that is, viewing events or objects located beyond naked eyesight, or even lost articles and people. It can be used to help or to pray with a recently liberated spirit, such as happens with death, and it can be used to place hexes on enemies, as a research tool, or for soul-walking.

It is pretty clear that tobacco, like its near relations cocaine, opium, henbane, and so forth, is dangerous, and that this danger is its appeal to sacramentalists. Even in the sixteenth and seventeenth centuries, Europeans made much of tobacco's medicinal properties. Another queen, Catherine the Great of Russia, is said to have so lauded its healing value that the craze for it among the nobility arose as a consequence. Its medicinal properties were recognized on both sides of the Atlantic soon enough, but on this side of the pond its sacred properties—that is, its power to alter this gross material world to specifications sent from the manito—was a driving motive for its use. In many Native societies, its use was reserved to special classes, or even individuals.

For instance, the great Moctezuma, emperor of the Aztec empire, enjoyed his smoke after dinner each night. His dinner was served to him on golden platters by scores of beautiful maidens—probably priestesses. The record says that this repast, which he took alone, featured an average of three hundred different dishes. (So much for the popular idea that Indians longed for the lifestyle of Europeans because they had never seen such luxury.) While the great emperor dined, he was entertained by tumblers and dancers, performing tricks and acrobatics, accompanied by wind instruments and soft percussion instruments. After the repast was cleared, the performers dismissed, the maidens gone, Moctezuma enjoyed a pipe or three.

Happily stoned, one presumes, he made off to bed, or perhaps to make offerings, to muse on the coming disaster (a word that means the stars are ill disposed), perhaps to pierce his tongue or penis with penance grass. Who knows? We know that the use of tobacco would dull the sharpness of physical pain, and should the royalty of Meso-America, the Andes, or kingdoms farther north have been afflicted by conditions such as ADD or ADHD,[29] manic depression, or other states of dysfunction (as presently characterized), or by states of high spiritual endowment (as then characterized), the sacred weed would, we know, have served to ease their minds.

While tobacco's sacred propensities didn't play well in London, Moscow, or Madrid, the plant got a lot of play for exactly those properties on two continents. From the Andes to the tundra, from sea to shining sea, tobacco rocked monarchy and monotheism, forcing both to drastic change, toppling both from the preeminent power place they had theretofore occupied. Definitely sacred stuff, it seems, to have so powerful and long term an effect. Not surprisingly, the sacred substance has been prohibited across all of the monotheistic world at one time or another, for one reason or another, since its wisps showed its presence in Europe and beyond.

Like the Indians of old, the Europeans heard about the goddess gift before they saw it. News of the exotic weed traveled with Columbus eastward, but it was a few years until it crossed the sea. When it did, its movement across all of Asia, including Europe, of course, and into Africa was swift and telling. One of the many ironies of the movement of this world-altering substance was its couriers: the priests who served an order that defined the plant as satanic transported it. Among the devout Catholics from Spain, the priests, like the soldiers and adventurers, were quickly habituated to the drug, although they used it in the form of snuff. The church seemed to see this mode of ingestion as preferable to smoking, largely because you could enjoy it during Mass and other stints in church without the disruption smoking it would cause. The Catholic priests, like the priests of Indian American empires, thought to keep this amazing stuff to their own class, but their use soon attracted the attention of their flocks, eventually causing prelates to issue orders that snuff use was prohibited before and during administration of sacraments, under penalty of excommunication.[30] Hard penalty indeed, but evidently it resulted in less public use of the substance. However, that could hardly affect the congregation in Lima, Peru, where the order was promulgated. The Native people there, descendants or thralls of the Incas, had long enjoyed the substance, and whether or not the aliens' "brown robes" indulged couldn't have had much effect on them, at least not then.

As near as I can tell, it hasn't had a particularly deleterious effect over the centuries, either, for the cradle of tobacco was also the cradle of cocaine, and the drugs combined were much in vogue among the indigenous people, from lowliest servant to great Inca, centuries before the coming of the cross-bearing bearded strangers in long skirts. It was also connected with corn and beans, and it should be said that none of the sacred plants was naturally occurring. They all had to be cultivated, developed from their wild form into plants that required settled agricultural communities for their optimum survival, leading to the development of sophisticated societies and complex civilizations.

It might be noted that the use of these natural consciousness-altering substances wreaked far less destruction on the indigenous peoples than that wrought by the Europeans. Lust for political and economic power, along with unyielding conviction that the values one subscribes to must be held by all, have done far greater damage than ever did a weed. It seems there are some demons that even sacred *apook* can't quell.

Conversely, the sacred weed has had a profound effect on the civilization that believed it was conquering and colonizing *apook aki,* tobacco country. Gately expresses perfectly the two-way, interactive nature of colonization:

> When one civilization absorbs another via conquest, it sometimes adopts its victims' culture to the extent that it is later hard to distinguish the victor from the vanquished. In such circumstances the vanquished entity may be said to have triumphed—through the perpetuation of its gods, its art, or its pleasures. The [Indians] may have been [all but] destroyed by the [Europeans] but their tobacco habits have been adopted since by the entire world.[31]

The connection of the four sacred plants—tobacco, corn, beans, and squash—with the "manito-ess," so to speak—the feminine intelligence of the implicate order—was not accidental. Over the continent the relation of settled communities and their orderliness, customs, and governing concepts were associated with feminine transcendent reality. By contrast, the agricultural communities of first Mesopotamia (Iraq), then Africa (Egypt, Nairobi, Sudan, and Ethiopia), and eventually Caana and Phoenicia (Syria, Lebanon, Israel/Palestine, and Jordan) were anciently associated with the masculine principle. In this rubric, the idea was that the seed is male, and the active principle; it is ejected into the female (earth, passive) and it grows itself. That construct can be said in Spanish—*se crece,* "it grows itself"—but not, interestingly, in English. The implied meaning is thus obscured for English speakers. So, in cultures that derive from Mesopotamia, which includes most of western, southern, and northern Europe, a plant grows itself, and thus the female is unreal, not of value other than as beast or slave is.

Tobacco, on the other hand, *divina planta,* and the world from which it springs, is grounded in quite another concept of how human institutions and power come into play and that tobacco, as manito, maintains. On the east side of the Atlantic, patrimony is of paramount significance. It is the home of the explicate order, and John Rolfe was a well-versed neighborhood lad. On the west side matrimony is queen. As our dear Matoaka and her people, however her story is reconstituted for patrimonial minds, well knew. She enchanted the already ensnared young English alchemist, and from her seat as his beloved spouse she tutored him in the sacred specifics of tobacco growing.

This is not to say, because we cannot say, historiograhically speaking, that Rolfe became a priest of the *tsenacommacah.* It is also not to say he didn't. The record shows that he was a pious Puritan, or so his letters and reports seem to prove, but as my granddaddy used to say, "Talk's cheap." Or, to paraphrase Samuel Goldwyn, documented evidence isn't worth the paper it's written on. We do know that many if not all the documents were tailored for their audiences; in a climate where people were being burned after horrendous torture, their families often suffering along with them, it is not likely that anyone would willingly admit he was dancing, not to mention sleeping, with the "divill."

Walter Ralegh, once the queen's courtier, had displeased her and found himself confined to the Tower of London. Freed within a year, he went on to great adventures that included the founding of Roanoke and with it gaining a royal charter for the land of Virginia. It was that charter that authorized the later expedition to Jamestown at the time when Pocahontas met her fate.

After James ascended the throne of England, Sir Walter Ralegh again gained royal displeasure and was accused of treason. Sentenced to death, he gained a reprieve and was then sentenced to the Tower of London, where he spent twelve years, writing and meeting with visitors and making his plans, smoking mightily the while. He was freed in 1616. Ralegh was believed by many at that time and since to have been the man to bring tobacco to England. He had tried cultivating it at his estate in Ireland, with little luck; his potato enterprise, however, was more successful. Other candidates for the honor of being the Englishman who brought tobacco to the English were the adventurer-pirates Sir Francis Drake and Sir John Hawkins. These three are the most famous of the men who operated on the high seas, enriching England's coffers and impoverishing her diplomatic standing in Europe. They attacked Spanish galleons, boarding them when they triumphed and making off with the cargo. Often that cargo was more tobacco than gold, and thus smoking, snuff, and a new kind of consciousness made its way to their homeland. By the time Jamestown was under way, Ralegh was languishing in the dreaded Tower of London as the poor addict Rolfe proceeded with his experiments.

James I was known to have ordered and participated in the torture and burning of thousands of people accused of trucking with the devil, whom James feared more that he feared Spain. Then again, Sir Dale, the fanatical tyrant who enforced his rule on the spiritual as well as the political front, did not countenance tobacco growing. He was a devoted Puritan, and believed profoundly in the role Satan could play in the life of the unwary. Self-appointed judge, jury, and executioner, Dale acted in such a way as to make it unlikely that a young soldier under his command would go against him.

It was under the pall of such dangerous circumstances, at risk of terrible torture and burning—even death—that the young couple took to the task of nurturing a dream that comes as smoke. In danger from their Puritan guardians, brutal Dale and pious Whitaker, the couple plied their craft. Should their activities have been discovered by their keepers, their fate, and the fate of the world, might have been very different. Nevertheless, with the quiet courage that distinguished her through her short life, Rebecca the priestess-magician, with her husband, John the alchemist–tobacco devotee, married, worked, and had quite a family. Along with the several thousand people who claim descent from the pair both here and in England, our globalizing world—this entire system of shared values and common assumption about the nature of reality and human roles in it—has grown out of their mating.

In a slightly earlier era, John the alchemist would have devoted his work to arcane methodologies current at Elizabeth's court and thriving, albeit moving out of courtly circles, among those familiar to the Temple, Gray's Inn, and the upper rooms at pubs frequented by the initiates. Among them were men such as Christopher Wren, who designed that singular edifice St. Paul's Cathedral; Ben Jonson, who entertained the court with his masques; Sir Walter Ralegh and company (before his arrest, of course); Francis Bacon and his sons and nephews; William Penn, who went west and danced in the woods with the locals, and named his lands Pennsylvania and the major settlement, City of Love; and William Shakespeare, the immortal bard. Mossiker suggests that among these intellectual pub-crawlers were William Stuckeley and John Smith. In an age when specialization wasn't prized, when men of wit and learning were a major power center in their own right, intellectuals, fighters, lovers, captains, adventurers, and captains of industry were habitués of certain neighborhoods in teeming London. As a subculture, they enjoyed certain entertainments, such as theater and hours at the coffeehouses, where they discussed heady discoveries, ancient manuscripts, anatomy and necrophilia (which those who dissected corpses, presumably for medical reasons, were charged with), architecture, and pamphleteering. They participated in certain men's societies, and mastered certain spiritual disciplines that went far beyond those of pulpit and Sunday service.

They all knew each other, of course. London society then, as now, was a merrily incestuous group. Little that transpired there or at other major centers in England, Scotland, or across the channel was missed by the London community's observations. The Virginia Company's settlements were lightly populated. Eyes and ears were everywhere, for intrigue was the lifeblood of the court and of mercantile society. Gaining favor with the court, or with the leading divines, was attractive to those who used favor to advance their interests.

Despite their peril, the couple persevered, producing a healthy crop of a new tobacco that would soon have the world afire.

Perhaps they knew themselves safe enough, at least for the purpose they had in mind. For whether Rolfe subscribed to the heathenish worldview of his wife and her people, he had a dream: he was going to make Virginia, via tobacco, a going concern, and he would not be deflected from his goal. Rebecca, for her part, had the Tradition, her training, the Great Council's advice, and her manito all making her path clear to her, and strengthening her will to realize that which gave her life, the great *powwaw,* together-we-Dream-Vision, of the *manitowinini,* "the people of the dream." Nothing less than a great world-renewal ceremony was transpiring. It seems that Pocahontas and her mate were the priests presiding over the rites.

POCAHONTAS

(only portrait done in her lifetime, 1616)

Engraving by Simon Van de Passe *(courtesy of the National Gallery, Smithsonian Institution/Art Resource, NY)*

Tapacoh / At the End of the Day

For Josh and I fancied ourselves among the New Men of
England. . . . We understood well the most important lesson
of the time—namely, that the real wealth of Englishmen
would henceforth derive not from the actual doing and
making of Things, not from the owning and exploiting of
Lands, but rather from the buying and selling of *Prospects*
and *Risks.*

—Gary Krist,
Extravagance

When they will not give a doit to relieve a lame beggar,
they will lay out ten to see a dead Indian.

—William Shakespeare,
The Tempest

As we begin our consideration of the last act of Pocahontas's life, it becomes possible to discern patterns of meaning that reveal her identity and purpose. The puzzle of how the life of a significant woman, which is a central theme in the Old Tradition on both sides of the world, takes on meaning beyond its own time and context emerges. We begin to see why the question "What time is it?" points to a deeper understanding that the times, cosmically speaking, determine events, placing them in many ways beyond the reach of human agency. Or, more completely, we begin to understand that human agency can operate only within a larger pattern that includes the motions of the planets and stars, the seasons and cycles of the earth and moon, and the currents of the Great Mystery, the *manito aki*. It is this understanding, whose entirety is encompassed in the Oral Tradition and how the events of Pocahontas's life and times embrace both, that emerges.

As can best be summarized within the confines of the linear, it can be argued that "Matoaka," our heroine's true personal name, suggests manito-woman; as said earlier, *matoak* means "white bird." However, the first syllable, *ma,* always signifies a connection or reference to manito. Because the syllable *ni* is absent, it can be assumed that something less than or somewhat other than manito is signified, although the word construction of Algonquin is complex enough that it's hard to say exactly why that particular syllable is missing when both *ma* and *to,* two of three of the required syllables, are present. Given the connection of White Bird Mataok with the creation story about Sky Woman and the waterfowl, which could have referred to a swan, her given name may have signified her character and destiny. The swan feather is the emblem of a Beloved Woman, and Pocahontas/ Matoaka was never pictured without one, even in her English portraits. Sky Woman or Flower (later to be known as Nikomis, which means "grandmother") was the supernatural being whose *powa* provided the energy or force field for a physical place on the planet to form, and it was because the waterfowl—swans—caught her as she fell through the void that she was able to create the *manito aki,* which in its turn gave rise to the material place where we live. The sequence or configuration strongly implies that a certain kind of female being coupled with white feathers equals great metaphysical power. Matoaka was such a one, as her astonishing accomplishments in the course of her short life testify. Alas, our most recent Sky Woman "avatar," so to speak, did not live long enough to meet her grandchildren, but they number in the tens of thousands and live all over the world.

Even the name "Pocahontas" shows that she was recognized as a nascent *powa* woman. This name might allude to various aspects of Powhatan life that would have been accessible to a child. Usually a child gets a certain nickname because of something the little one does, says, or resembles in character or appearance. Matoaka's child name might have been related to the Powhatan word for "beaver," *pohkewuh.* This possibility raises some interesting avenues of thought, for beaver, or "earth diver," was, in some versions of the Sky Woman story, the being who gave his life to bring earth up from deep in the void. It was upon this bit of mud that Sky Woman was lain by the waterfowl guardians who saved her. In a related possibility, a Lenape word, *amochk,* also means "beaver." Lenape is the language of an Algonquin group who traditionally lived in what is now Delaware, and is considered by linguists to be closely related to Powhatan. So closely, in fact, that some, such as Native scholar Jack Forbes, consider them dialects of the same language group, which makes them close to mutually intelligible. Pocahontas's secret name, Amonute, could conceivably be related to this Lenape word, particularly as sacred or secret words, used in ceremonies, often derive from an older or "proto-" version of the language as it is commonly used. Lenape and Powhatan clearly descend from the same protolanguage, so what was common in one dialect might have been all but lost in the other.

There is another word that might be the source of her nickname: *pocohaac.* Translated variously as "awl," "needle," "penis," "pin," "bodkin," and "pestle," the word, or concept, points to complex relationships between women's power realm and men's. The various meanings of *pocohaac* are all women's implements, with a revealing twist on Freudian associations, and all are connected to magical as well as household activities. *Pocohaac* comes as close as any word to her child name. *Pocohaac* is not, as it might seem at first glance, a house-keeping implement that signals a woman's eternal state of drudgery. In ancient days in Mesopotamia and elsewhere, these female implements, the world's first tools and therefore the wellspring of technology, were used in magical and ritual work, first by female elders, later by men.[1] Eventually these same implements, and in particular the superior creative, mental, and social-bonding power they represented, became the staffs of shepherds, the scepters of kings, and the wands of wizards. As in early Middle Eastern and European societies, the peoples of the Americas recognized the *powa* inherent and implicit in the original tool; *pocohaac* contained in word and fact the history of the human world. Male-appearing implements, found useful for improving the quality of life of the entire community, when wielded by women are highly significant: a person perceived as *pocohaac*—awl, needle, pin, or penis—when that person is a girl child, is perceived as peculiar and, because anomalous, significantly spirit haunted, spirit driven, spirit blessed. Whether *pocohaac* or *pohkewuh,* Pocahontas is a name that implies a connection to mystical matters, particularly those with direct bearing on woman-centered ritual tradition. In both of these words there is a bit of a trickster element: both are creative in function, closely related to *powa* in significance, and a bit off the "straight and narrow" in action. Perhaps it is this undercurrent that led those who acted as informants to the English to translate her name as "wanton," "frisky," or "mischief." There's no reason to suppose that Indian humor has changed more than most other aspects of Native thought and expression over the centuries.

Her path was clear from the moment she was born, perhaps before her birth. Some of the old-time people could see generations ahead, and know exactly what child would be born, what their destiny, even their size, shape, and countenance would be. In a case like that of Pocahontas, given the trajectory of her life and its far-reaching significance, her conception would have signaled a "heads-up" to many an older Dreamer. That Wahunsenacawh served as Powhatan, Principal Dreamer, during her entire time among the English, dying not long after she did, speaks volumes about how his great power came to him, and in whose interest it was held.

"Rebecca," the Christian name that Pocahontas took upon her initiation into the Christian universe, is a story in itself, and follows a seemingly serendipitous path as her other names do, but there are few accidents in the life of a familiar of the manito; she may have chosen it for herself, being told of Rebecca, who according to Scripture was the mother of another nation, Israel. Finally, "Amonute," her secret name, makes a certain aspect of her unknown—that is, mysterious, hidden, veiled, sacrosanct. In this guise her identity is that of the High Priestess known in Enochian magical systems and more popularly through the Tarot.

Taken together, these four names—Matoaka, Pocahontas, Amonute, and Rebecca—provide a four-dimensional identification of our heroine. As she is an Indian woman, and in the Native world four is the basic sacred number, perhaps her acquisition of four identities or aspects is one more indicator of her stature as a Medicine Woman. Four and seven (which is the other widespread sacred number) mark the levels of mastery within the *midéwewin*, the Medicine Lodge. It is usually understood among the community that there are four levels of mastery in the Medicine Lodge; it is less well known that there are an added three. One who masters these seven levels is at a level that would be recognized as "magus" in European pagan spiritual lore. Perhaps in the Powhatan universe, such were the most influential leaders, the *mamanantowick,* male, and *mamanantoskaa,* female, one of whom became so adept at walking in the *manito aki* at will that he or she was known as Powhatan.

Considering matters that are obscure if not necessarily arcane, there is the matter of the alchemical wedding: where it was performed, and by whom. It is believed that the great event took place at Jamestown, presided over by Rolfe's good buddy and associate of the other faction headed by Sir Thomas Gates, the Reverend Richard Buke. But there's the consideration that the Reverend Alexander Whitaker was responsible for guiding the heathen princess into the fold, under the careful sponsorship of Sir Thomas Dale, who had removed said heathen from Jamestown and the influence of both Richard Buke and Sir Thomas Gates, the better to be the one responsible for her education and conversion. The site currently offers a brochure stating that the foundations of the old church—the rest of the ruin had been erected some time after the edifice standing at the time of the nuptials—were the same that supported the original church where Princess Pocahontas married John Rolfe. Of course she was no longer Pocahontas, having been baptized Rebecca prior to the marriage. While legend is a powerful force in its own right, there seems to be no document that supports either parish—Henrico or Jamestown—as the site of the nuptials. The famous painting of America's first mixed-race marriage that hangs in the rotunda of the Capitol Building in Washington, D.C., seems to favor the church at the original James Fort.

Another of the historical mysteries that surround the Rolfe relationship from courtship to householding is the location of their original holding. There are a number of candidates for this historical honor and the tourist trade it might entice. There are three possibilities: spreads located at Varina, near Henrico; Mulberry Island, quite a distance away; and Hog's Island, likewise. Varina is the popular favorite, and is thus accorded the honor of being the official location of the newlyweds.

The lovers settled down on a piece of land they named Varina. According to legend, it had been a gift to Rebecca from Wahunsenacawh, although it might well have been him along with *weroanskaas* and *weroances* or the *matchacómoco,* the Great Council, or the council of whatever tribe called it home. Located outside Henrico, it was named in honor of the tobacco that Rolfe, Rebecca, and her Indian companions developed. They combined the Caribbean tobacco that Rolfe had imported, stolen, or traded for via some drug-smuggling route of the era with Native techniques for cultivation and curing and the most successful of the numerous experiments with seeds and seedlings Rolfe had conducted. One can almost see the wheels of a gifted marketer at work in this happy choice of name. The first crop planted for export by the English would be wildly successful—a true wonder when one considers the combination that went into its production. Like the union of John Rolfe and Rebecca, which produced a mixed-blood boy named Thomas, whose descendants would become as numerous as the grains of sand in the desert, the union of Powhatan and European methods of developing, cultivating, curing, and marketing would produce a nation whose wealth would become the greatest the world had ever seen.

The young lovers Becky and Johnny happily-every-aftered on their land, wherever it was, he experimenting carefully with his golden horde, she swelling with maternal satisfaction, the fruits of love's labor simultaneously ripening within her and the good earth. Evidently the Mother within and without the Rolfe homestead was hospitable enough for the little family who would found a nation on the happy mating of human, plant, and timing. The fourth part of the mix, as ever, remains hidden, for it belongs to manito, the mysterious.

It seems unlikely to me that Rolfe had any idea of what he would wright (isn't *wright* the present tense of *wrought?*). But Rolfe, whoever he was, was the first modern professional businessman and, as such, the harbinger of a class that hovered just offstage awaiting its time before the footlights of history. The Pocahontas narrative featured three prototypes. The first is the American heroine, a woman who is strong, self-determined, brave, independent, and brilliant, gifted in affairs of state and possessing control over her sexuality. Pocahontas may have been the prototype for a type of heroine largely confined to the American scene, both in real life and in popular culture. The second, the hustler-adventurer with strong competitive instincts, has organizational flair, powerful leadership capacity, and dislike of institutional authority. This type has a strong ability for public relations, and thus John Smith becomes the prototype of the American hero: he is the template for Deerslayer, Shane, Rooster Cogburn, and Captain Kirk as well as for Superman, Batman, and Captain Marvel. The third, the primarily professional entrepreneur, is oriented toward scientific research and development. This is the prototype for the American man: politically savvy and able to negotiate with established institutions. In some instances this type is reflective and full of personal doubt, yet he forges his way to financial success. It is he who is the subject of literature—of works such as Arthur Miller's *Death of a Salesman* or Sinclair Lewis's *Babbitt*—men who seem weak because they aren't adventurers.

This template is the hero who signifies conservative American values: piety, peace, progress, and prosperity. Of this trilogy, the Rolfe-type has prevailed by and large. Significantly, his ability to promote world-wide power is catapulted and buttressed by the independent woman, the helpmate of whom it is said, "Behind every powerful man there is a strong woman" or some such, and the adventurer-hero. The story of Pocahontas seems to tell us that while John Smith appealed to the *tsenacommacah* as their candidate to be the first English-Indian, Pocahontas's selection of a less glamorous but much more devoted Englishman produced peace for the *tsenacommacah,* albeit fleeting.

Perhaps, to continue in a literary vein, John Rolfe and Rebecca/Pocahontas were star-crossed lovers, fated to die before the fruits of their passion had had time to do more than bud. One considers doomed lovers such as Antony and Cleopatra, Tristan and Isolde, or even Arthur and Guinevere. On the other hand, perhaps these star-crossed lovers were so primally American that no prototype for their doomed relationship exists.

Presumably the young couple spent quite a bit of their time at church or wherever Alexander Whitaker held his prayer meetings. There was also much to-ing and fro-ing of *tsenacommacah* members, the bride's "family"—which, to her, would mean her clan relatives, or perhaps fellow students or even mentors from Werowocomoco. One would give much for a record of the matters discussed among them. It is sure that they instructed Rolfe in tobacco cultivation in tidewater soil, the best fertilizers, planting patterns, and the like. It is equally likely that they returned to their clan and medicine buildings to con-fer with the tribal council, the elders, and spiritual adepts, male and female.

During those months, whatever plans the Powhatans had involved peace and harmony between and among the various groups in the area. During this "honeymoon"—not so much that of the couple, though that too, but that of the diverse communities—the English settlements boomed. The Indians waited, knowing this "Indian sum-mer" couldn't last; they had their ways of knowing what was likely to come; in addition to Rebecca, they had astrologers, soul-walkers, and advisers within the *manito aki* to rely on for intelligence. Rebecca/ Matoaka, growing ever more versed in English and the ways of English society, was refining the skills she would need during the next, and final, act of her role in the Great Play.

On June 12, 1616, the curtain went up. The Rolfes, like a frontier version of the holy family—young mother, older father, and small son—along with a retinue of Powhatan nurses, shamans, and medicine women, embarked on the difficult voyage east to London, where Lady Rebecca would be greeted by the cream of English society, feted, paraded, gossiped about, and finally—so some whisper— murdered. In that period, the English were aware that they were in a new era. The idea of modernity was much bruited about in pubs and salons, and the cognoscenti of the age, men such as Francis Bacon (1561–1626), the father of the scientific method, and Galileo Galilei (1564–1642), the father of modern astronomy, were turning the world, hitherto based in religion, in a new direction, with the great architect Sir Christopher Wren (1632–1723) and the father of modern physics, Sir Isaac Newton (1642–1727) soon to follow. It was a period in which both the globe, and the universe as well, were beginning to yield their secrets to men of the Eastern Hemisphere. The idea of the spherical earth revolving around the sun, an earth-shaking concept, opened minds and markets around the world. In this age of ferment, the ideas of the preceding era that were shaped on assumptions concerning the supernatural and the theological were being supplanted with ideas based on assumptions of the physical and material nature of reality. The implications of the swing between these ideological poles was new. The English of the early seventeenth century did not yet realize that their absolutist universe was morphing into a relativist one, that their absolutist society was on its way to republicanism and democratization, or that their religious worldview was on its way to scientific rationalism. Had they known, maybe they would have stayed home. Maybe . . . but probably not. What with the drive for more power, which created the European nations' need for more money, and the growing diversity in Christian religious sects as well as in information in general, a powerful instinct was aroused: curiosity.

As European society's rigid feudalistic rule fragmented, the thirst for new knowledge that this change engendered would have driven their voyages of discovery anyway. It was inevitable that they would come here. A major force driving them, other than the quest for national power greater than that held by Spain or the Roman church, was their belief that they could plant a Protestant colony of the Puritan kind far from the intrigues and complexities of Europe. This was a cause to which a number of influential merchants and clergymen such as Richard Crashaw, John Donne, and the Bishop of London, John King, were ardently devoted. They announced that their venture would enable them to put a bit in their "ancient enemies mouth." I'm not sure if they thought the Catholic Church was a horse, or if they were thinking in terms of inquisitional torture with that metaphor. What with the growing merchant class's need for markets, goods, and resources and the Puritan church's devotion to making its mark on what it termed the "salvages" in that far land, the drive for colonization was reaching critical mass.

Meanwhile, King James I was singularly indifferent to these forces; he had wealth enough to indulge his queen in her lavish lifestyle, while pursuing luxurious tastes and expensive endeavors of his own. His children, he believed, were set to marry well among wealthy European potentates. He was the first king of the United Kingdom, of a land that was enjoying a span of peace and prosperity after decades of civil and international war.

FROM RIGID TO FLUID AND BACK:
TIMES OF CHANGE

It's fascinating to explore the ways in which the implicate order thrusts itself into the explicate order over time. One might read history as the interplay between the two, or among the varieties of waves that shape the currents that can roughly be grouped into two opposing tendencies of existence, one of which is the explicate—society as it exists at any given point in time when its rules are generally agreed upon and externalized. These rules become solidified then petrified. Existence doesn't seem to like absolute rest. The universe, scientists tell us, is not only spherical; it is in various degrees of acceleration. And, they add, there is more than one universe . . . which implies that the universe is part of a multiversity. The understanding that the universe is in constant flux and thus nothing remains fixed characterized the worldview of the *manitowinini* at the time that Europe bumped into them. This point of view directly conflicted with the absolutist mind-set that had governed society and thought for more than a thousand years. In a way, the Indian world institutionalized the concept of motion and change as the natural order, an idea that Puritanism sought to deny as intensely as the emerging sciences and free-market systems forced its acceptance. It is only because of this emerging dynamic in European thought and life that a woman such as Pocahontas could have the world-changing effect she had. The Indians knew this. It is because indigenous science recognized the fundamental significance of "time" (as in energy or thought-being currents outside of human control but functioning in chartable cycles) that the Powhatans could devise a strategy for the long run. Records kept in the Oral Tradition of tribal nations across the Americas demonstrate that the Indian strategists of the Times of Change and Renewal were planning far ahead; perhaps their plan extended over centuries, to the change of the millennium, common time, because the period from the late 1990s through the first decade of the twenty-first century is recognized in Indian Country as "the Seventh Generation," when the results of their efforts would become clear.

OKE VS. AHONE, MOVING VS. FIXED

The English in the tidewater area were aware of a particular being the Indians called Oke or Okee, as near as the English could make out. This entity seemed to the English to be a god to the indigenous people, and they recorded that effigies of it were used in battle, often carried before the advancing warriors and brandished at the enemy. The Algonquin of both the northern and southern sections of the eastern seaboard referred to these effigies as Oke (or Okee). The English understood that Oke was seen as a god by the locals, and took great delight in demonstrating that they, the English, could assault the effigy, smash and burn it, with impunity because it was only animal hide and wood.

What the beneficiaries of this crude instruction made of the weird behavior the aliens displayed is not recorded, but as so many questions remain unasked and unanswered, for the record, at least, we must surmise their response. William Penn, the English "founder" of Pennsylvania ("Penn's wood"), records an exchange that took place in the woods where he chose to settle. He asks one of the locals about the Oke, and is told that Oke is a god, one that is vengeful and dangerous. He must be placated in numerous ways so his negative or malicious effects might be deflected. They assured him that Oke wasn't the God, or *mannit,* which led Penn to ask why they prayed to Oke and not to Ahone. His informants explained that Ahone does not need to be attended to in any way, is not worshiped, because this deity is good and can do nothing but good as naturally as the air cannot be other than air, or as rivers flow as they do. What is most striking about this exchange is the uncanny resemblance it bears to the current Christian understanding of the sacred: God is the ever good deity, while Satan (or Lucifer) is vicious. There the resemblance between the two sides of the discussion diverges, of course, for no Christian would seek to seduce, placate, or entice Satan to act kindly on the quester's behalf, while every Christian would, and did constantly, beg God for help, forgiveness, blessing, and salvation.

In Oke and Ahone, it seems that William Penn stumbled on the sacred twins of Algonquin (as well as Haudenosaunee) myth. They are the children of Winona, daughter of Sky Woman. They are the twins who make the earth, Turtle Island, ready for human habitation. Ahone, who is the primary maker, ensures that the entire biosphere, including geological features, climate, and plant and animal life, works to the advantage of the humans (who are yet to come). Oke, who envies his brother's ability to create an entire planetary system, tries to make things also, but everything he makes is perverse; that is, it complicates human endeavor to the point of destructiveness. Thus, Ahone creates the great rivers, upon which the people depend for travel, to run both ways so that one can catch the current and go with it to one's destination. There is no "up" stream or "down" stream. Similarly, corn and tobacco germinate and mature without human intervention, and always thrive all on their own. But Oke's rivers run only one way, so that people must strain against the current to reach any destination "upstream" from their starting point. It is Oke's idea to make corn and tobacco grow under conditions that require human cultivation, and he adds cycles of drought, extreme winter weather, and exceeding rainfall to complicate matters further. It seems that Winona gave birth to the principle of polarity, without which, of course, the material world we know would not exist.

Lives that have the impact that Pocahontas's life had are generally seen, traditionally, as reiterations of powerful lives enshrined in the Oral Tradition. Seen this way, perhaps part of Pocahontas's mission was to discover the English versions of Ahone and Oke. As a recently indoctrinated and baptized Christian, she would have been instructed about God and Lucifer, of course. She may have found herself as perplexed, if not as pained, as her contemporary, the English Puritan divine and metaphysical poet nonpareil John Donne. The dilemma faced by seventeenth-century Puritans is well considered in the ninth of Donne's "Holy Sonnets":

> *If poysonous mineralls, and if that tree,*
> *Whose fruit threw death on else immortall us*
> *If lecherous goats, if serpents envious*
> *Cannot be damn'd; Alas; why should I bee?*
> *Why should intent or reason, borne in mee,*
> *Make sinnes, else equall, in mee more heinous?*
> *And mercy being easie, and glorious*
> *To God; in his sterne wrath, why threatens hee?*
> *But who am I, that dare dispute with thee*
> *O God? Oh! of thine onely worthy blood,*
> *And my teares, make a heavenly Lethean flood,*

And drowne in it my sinnes black memorie;
That thou remember them, some claime as debt,
I thinke it mercy if thou wilt forget.[2]

Had a similar plaint been penned by a seventeenth-century Powhatan, perhaps it would have been addressed to Oke rather than Ahone. Be that as it may, it might have seemed to Pocahontas that God had a strange attitude toward hapless creatures deemed capable of "reason." She might well have assumed that the manito whom the Reverend Richard Buke and her husband called "God" was the English equivalent of Oke. For as one of his Native sources earnestly informed William Penn, the Algonquins made much of placating Oke. They didn't have to do anything to please Ahone, because he was always joyous and pleased with all that existed. Given the system in which she had been educated, the Christian dichotomy between God and Lucifer must have seemed to her Algonquin mind exactly like that between Ahone and Oke, and thus not at all difficult to grasp. The Powhatans informed the English that one need not pray to Ahone or make any efforts to attract Ahone's beneficence, for he was all good and would do nothing other than good to and for them.

Donne's anguished meditation bespeaks a mind torn between the two ideas of God: Is the divinity an all-loving being? If so, why does he love only the other animals, even the plants, but not humans, not his loyal servant, Donne? It's a conundrum that the Powhatans along with their northern Algonquin cousins had resolved by making two gods: one who had malevolent intent, the other only beneficent. Christians of the time—and earlier—had constructed a similar supernatural universe, positing a god who is all good and a devil who is evil; from the former comes eternal life; from the other, hellfire. Until the Puritan and allied sects' reformation of Christianity, Christians believed that the act of choosing good over evil had an effect on one's place in the afterlife. However, the Puritans believed that one was predestined for one or the other and that personal acts had no effect on the outcome. It was up to God to decide, and up to humans to accept. Later, as the American colonies got more established, pastors and divines would thunder from their pulpits about the fate of "Sinners in the Hands of an Angry God."[3]

The Powhatans seem to have arrived at a parallel understanding of the workings of the sacred universe, but their solution was different. They attempted to utilize and placate Oke, the negative force, for the community's good, believing that if humans could overcome Oke's negative influence, goodness would necessarily prevail—an idea that would find its way into English and American thought in the nineteenth century with the rise of Romanticism. For the Algonquin, then, human activity was central to human happiness, here and hereafter. But by Pocahontas's time, seeing the greater power, *powa,* available to the English, the Powhatans were rethinking their cosmogony, as would most of the other Native Nations in ensuing centuries. If the Christian god, they reasoned, made it possible for his priests, conjurers, and magicians to channel power that enabled them to have such amazing possessions—muskets, compasses, great sailing ships, glass and pottery containers, and the like—his *powa* must be far greater than Oke's. They determined to discover how that magic, that English *powa,* worked, with a view toward adapting it to their situation.

Thus Pocahontas, in all her dealings with the English, labored to discover the workings of this powerful deity and his priests. She would then channel the information back to the council of shamans and priests, the elders of the Medicine Dance Lodge, and they would do whatever was necessary. She had to be a highly developed adept herself, or she could not have been sent out into the realm of this Kitchee Manito, this Great Manito, with any hope of return.

In another "Holy Sonnet," Donne informs death that it isn't so big: the sleep it brings is easily replicated by drugs—which at the time were sometimes referred to by poets as Lethe, the Greek river of forgetfulness. "Thou art slave to Fate, / Chance, kings, and desperate men," the poem accuses. "And dost with poyson, warre, and sicknesse dwell." Even more damming, he charges death with existential inadequacy, since "poppie, or charmes can make us sleepe as well, And better than thy stroake." So, he finishes triumphantly, "why swell'st thou then?"[4]

AMERICAN LETHE

John Rolfe and the other gentlemen addicted to *uppowoc-manitos* (to-bacco manito), to coin a term, would have approved of the poet's conclusion there. The verse makes reference to the all too ready availability of the drug at the time, a fact that no doubt fueled the burgeoning market economy as well as the truculent peace between Spain and England and the unfavorable judgment of godly men. Assuming that the English reports were accurate: the Algonquins, like many other indigenous people of the Americas, geared their sacred practices to the local plants with sacred characteristics by offering copious amounts of the cured and especially prepared substance whenever an occasion demanded. Such occasions were common: any dangerous or potentially dangerous situation saw the bestowal of *apook* to the particular manito involved. It might be fire, water, snow, or the English that were honored, and to whom pleas for benevolence were addressed, but the offering was made to increase harmony. As centuries of research demonstrates, tobacco was the most widely used sacred substance. It was used to honor and appeal to the kinder side of manito like the sea, lakes, rivers, and ponds. It was given to fire dedicated to community or religious purposes; it accompanied endeavors ranging from planting and harvesting, remaking and initiation, birth and burial, to any other occasion designated as situated on the threshold between the mortal and the supernatural worlds.

At the time Pocahontas and her retinue arrived there, England and Spain were in the midst of a seventeenth-century version of the Cold War. Their relations had a profound effect on her fortunes, for her husband and the entire "Virginia" (as the seventeenth-century English thought of what we call America) enterprise had their money and dreams of financial independence for a Puritan church pinned on the success of their tobacco plantation enterprise. Concepts such as "cash flow," which, along with "balance of trade," was a new item in the growing lexicon of emergent capitalism, and concerns of raising financial backing, were of intense moment for certain investors in their struggle for spiritual dominance. These spiritually focused men adroitly used privateers (the current euphemism for pirates), market espionage, and military buildups in pursuit of their goals, providing some of the earliest models for how best to comport business on a multinational scale. The northern European nations, including parts of what is now Germany and England, were hotbeds of religious fervor that often spilled over into bloody wars. In France the same struggle was also raging, albeit with different results—for that century and the next, at any rate. The religious ferment occasioned by a variety of reformist movements and shifts in national and international affairs was reaching fever pitch, and the more overt battles—like the defeat of the attempted Spanish Catholic invasion of Protestant England that had failed early in the previous monarch's reign—and the ascension to the thrones of both England and Scotland by James I made those two nations firmly Protestant.

However, James favored a monarch-headed church, an idea that Puritans such as the financial backers of John Rolfe's tobacco enterprise loathed. By the same token, as "Counterblaste to Tobacco," his detailed diatribe against the use of tobacco, shows, the king heartily loathed tobacco, believing that snuff, pipe tobacco, and "chaw" were the embodiment of all he feared—his overthrow and execution by the witches and savage heathens arrayed against him. Perhaps James I was prescient—or haunted: such a fate befell his son and heir, Charles I, as it had James's mother, Mary, Queen of Scots. Mary had been beheaded by England's first Protestant monarch, Elizabeth I. Elizabeth ascended the throne a Protestant, or at least openly advocating Protestantism and repressing Catholicism, while her father, Henry VIII, had merely denied the authority of Rome over his marital choices. Charles I was beheaded during the final thrust for Puritan dominance in English politics in 1649 at Whitehall, under the rule of the first commoner head of state, the Puritan politician, military leader, and Lord Protector, Oliver Cromwell.

By the end of Elizabeth's reign, whether egalitarian or authoritarian, feudal or free of the rule of the vassalage system, the English church was irretrievably non-Papist. In Spain, on the other hand, a very different situation applied. The Spanish monarchy owed its allegiance, including a not considerable portion of its victories in the preceding century, to the Holy Roman Empire, located a short sail away in Rome. (Although, to be fair, Italian historians hold that Italy was dominated during that period by Spain.) Whatever, England wanted a major seat at the table of world power, an entire new world of greed and grasp, and it wanted its place unobstructed by its earlier vassalage relationship to the Papacy. With Henry VIII's defiance of Rome, England had landed the first blow against the Holy Roman Empire; in defying Spain by settling lands Spain had marked as its own (i.e., North America), England landed a further blow. While not a knockout, the settlement at Jamestown and soon after at Plymouth, coupled with Rolfe's economic strike against Spanish hegemony and the Puritan conspirators' financial investment in the new world they intended to call their home, would eventually result in the decline of Spain and Rome as world powers and the rise of Great Britain to possession of an economic empire upon which the sun never set. The confluence of events and personalities during an era that stretched from Elizabeth to Victoria also saw Indian America, Turtle Island, transformed into a cash cow (or maybe a goose laying golden eggs) for two major powers: Spain and England.

England's victory against the Spanish armada in the mid–sixteenth century had set the tone for their future. And to their good fortune, John Rolfe took up the cause of joining Turtle Island to England, the *manito aki* to Faerie, when John Smith ran away. Rather than John Smith, the working-class glamorous adventurer, it was John Rolfe the gentleman alchemist-planter who won the hand of Pocahontas, princess royal of the Empire of Apook. It was Rolfe, what with one thing and another, who chose the brand name Orinoco, a name redolent at the time with intimations of El Dorado, that mythic land of human immortality described by Walter Ralegh. In 1616 representatives from Virginia Colony, Jamestown, set sail for England under the leadership of Marshal Dale, with one princess, a model of savage heathen transformed into civilized Christian lady, aboard. They hoped to raise additional funds to keep their venture afloat a while longer, and for that reason they carried twenty thousand pounds of Virginia's finest, bred in northern shores and blessed by priests and priestesses of a northerly tongue and clime. With Pocahontas, Indian paragon of missionary zeal and cash crop, indicator of solid investment opportunity, firmly in tow, Sir Dale must have anticipated a warm welcome and highly successful outcome out of his meetings with the representatives of City of London Company, parent company of the Virginia Company. There's little doubt that he was as prepared for intricate maneuvering as any upwardly mobile corporate executive. That he was not, perhaps, as adroit at corporate maneuvers as he had hoped seems to become evident in the end, when he did not settle in Virginia as planned but was sent to another branch in 1616. But whether he was fully prepared for the court's warm reception and the company's rapidly cooling one had yet to be seen.

The aptly named *Treasurer* set sail in 1616, carrying its precious cargo: Lady Rebecca, her husband, and their son, Thomas, and her dozen or so attendants. In the mind of the rising venturer John Rolfe, the memory of the shipwreck of the doomed *Sea Venture* towered fearfully. Nor was he the only one with difficult memories triggered by being aboard that ship, constructed as it was from wood salvaged from the *Sea Venture* and, like it, captained by Samuel Argall. The *Treasurer* was the ship Pocahontas had boarded, innocently or not, when she sailed downstream to James Fort, and Argall had been the captain on that journey as well.

That sailing, however star-fated, was on a river; sailing the ocean blue was another matter entirely. It took a great deal of courage to sail in those perilous times, and the code of the sea stipulated that there would be no "devillish pursuits" aboard ship. Among forbidden activities were gambling, profanity, and ribaldry. While this code, known as the Code of Oleron, did not seem to specifically forbid carrying "devillish" heathens, the Englishmen aboard might well have been more apprehensive than usual on this journey. Among Lady Rebecca's party—none of whom had converted to Christianity—was a heathen priest of no little ability. This was Uttamatamakin (a.k.a. Tomakin, Tomocomo, or Tomo), husband of Rebecca's "half-sister" Matachanna and chief among her co-agents. A council representative (whether for the overall *tsenacommacah* or his own tribal council is

unclear), he painted his face and body and wore Native dress—a decorated scalp lock, a breechcloth ornamented with an animal's head
and tail, and a fur mantle. Also part of the royal party was another half
sister—or clan relative—along with three other Powhatan women
acting as servants, and four other Powhatan men. The retinue had
been organized by Sir Thomas Dale to enhance Pocahontas's status as
a woman of royal birth.[5]

It is unlikely that Dale realized he was shipping a party of spies
into his home territory. But it seems clear enough that this was indeed
the case, though the party's efficacy as intelligence agents would have
been ruined had the English been aware that the Native people were
practiced at political strategies. It was in the interest of the *tsenacommacah* to encourage the English belief in Powhatan ignorance and English intellectual and military superiority. One can't help but wonder
how they managed to convince the English of these illusions, given
the Jamestown contingent's chronic inability to survive the smallest
natural crisis, and given the Indians' clear military superiority. However, Smith's journals themselves clearly prove the human skill at self-
delusion. In the face of all evidence to the contrary, Smith continued
to believe himself greater than Indian and English elite alike, to see
himself the innocent victim of English and Indian machinations.
Thus, he continued to proclaim the malevolence and incompetence
of his enemies, Native or home boy, until his dying day. He did, it
seems, attempt to make some restitution for his own misbegotten
deeds. Writing at length about Pocahontas several years after her
death, when he was near his own death, he wrote that Pocahontas
had been one of the most important women in his life.

In actuality, Lady Rebecca and members of what the English believed to be her family were on a mission organized by the Powhatan
matchacómoco—Wahunsenacawh no doubt the most powerful among
them. Similarly, Uttamatamakin, identified by English correspondents
as Lady Rebecca's "father-councilor," was sent for much the same
reason. His cover: that he was the husband of Pocahontas's sister
Matachanna and also escort and counselor to the Powhatan princess,
Lady Rebecca Rolfe. Likewise, every member of the party had an of-

ficial identity—one that was likely close enough to what the English had been told about them, but missing one salient fact: they were all highly skilled shamans, *quioccaska,* and/or medicine people (i.e., priests and priestesses).

In the Powhatan world many religious ceremonies were gender-specific. Women had their rites, men theirs, and each group had information and protocols that were unknown to the other. Reasonably, the Powhatans would have assumed that the English had a similar structure—for humans generally expect that other societies will be based on highly similar, if not identical, assumptions. Thus they would send one adept from each group; Pocahontas the Beloved Woman, priestess and adept, and Uttamatamakin, priest and council member, himself a highly trained adept.

Uttamatamakin (Tomococo) was married to Matachanna, who was said to be Pocahontas's sister. Apparently, this conclusion comes from Smith's note that Tomococo was married to "one of Powhatan's daughters." However, as it is the greater likelihood that the relationship described as paternal-child is not biological, Matachanna and Rebecca were not necessarily biological sisters. It is more likely that they were sister-adepts, rather than clan or blood kin, which presents another picture. Rather than the lady and her baby's nurse, Pocahontas and Matachanna are two priestesses who between them, and with the added *powa* of the other women accompanying them, were able to evoke powerful forces. With Uttamatamakin, himself a priest of considerable ability, the party formed a full ceremonial complement, enabling them to engage in sacred work the results of which surround us today.

POWHATAN AMBASSADOR OR SHILL?

Lady Rebecca, medicine woman, spy, and diplomat, was kept fully occupied meeting and entertaining the major figures of the great city. Both there and on board the *Treasurer,* unseen forces swirled around the lady, a number of them seemingly political and economic, ordinary in a burgeoning market economy, while others were of a more spirit-moved kind. The two realms were brought into a condition of interfacing by her espionage and *powa,* magical efforts on behalf of the *manito aki.* While economics drove the Virginia Company to bring Pocahontas and her *powa* and *powagan* to England, far greater manito forces filled the winds that blew them across the ocean. There were two groups of investors keen to influence the outcome of this great adventure: the Virginia Company was out for a big financial return on its investment; while the manito were out for capturing a vibrant new field of dreams, an expanded *manito aki.* One of the passengers on the *Treasurer* was a Spaniard named Don Diego de Molina. This worthy had been captured by the English and held at James Fort, where he was kept, presumably in comfortable quarters, at the time Pocahontas herself was brought in. Whether they exchanged pleasantries or not, or whether each recognized the other as a fellow espionage agent, is not known. However, they may well have conversed. After all, Rolfe had somehow got hold of Bermuda gold, the tobacco seed that became the smooth, mellow blend that made the Virginia Company an extremely profitable venture. Perhaps de Molina was related to that mysterious event in some way, or at least suspected the theft. In any event, with results that only those directly involved in a conspiracy among them could know for sure, de Molina, Uttamatamakin, and John and Rebecca Rolfe were forced, by default, to spend a great deal of time together aboard that small vessel.

The intricacies of early seventeenth-century politics reveal much about the political realities of the time. Taken prisoner with de Molina was an Irishman named Francis Lambry. Lambry had piloted the doomed Spanish ship, and in accordance with his less-than-commoner status—no doubt compounded by his country of birth, which was then being occupied by the English—he was kept in rather more punitive conditions than those in which Don Diego suffered. On the public front, the English and the Spanish were adversaries, although their fight had taken on the complexion of a war of plunder, which had evolved from the state of armed combat it had earlier been. Around the time of Pocahontas's visit it was reported that "[t]he Spanish Ambassador went recently to the royal council to ask that Sir Walter Ralegh should not be allowed to go to the Indies [the

Caribbean]. He is to leave in two months with eight ships full of nobles, all well appointed, to acquire mines."[6] In 1616, on the eve of Lady Rebecca's tour, the Spanish ambassador to England brought suit against Sir Richard Bingley for piracy. Bingley had boarded two Spanish ships and taken the cargo, which included a variety of items from hides to redwoods. The Spanish, of course, had lost a sizable portion of Bermuda tobacco in the cargo hold. In the midst of these political tensions, the *Treasurer,* bearing riches both sacred and profane, set sail. As it turned out, it was sailing into an extremely fortunate future.

The Wall Street maven Peter Lynch offers an eye-opening rundown of investment ventures in the seventeenth century in chapter 1, "A Short History of Capitalism," of his best-selling guide to investment and business.[7]

Focusing on two major business endeavors of the time, the Dutch East and West Indies Companies and City of London Virginia Company, he provides a delightful excursion into the workings of a sibling company, the Virginia Plymouth Company. This corporation was split off of the Virginia Jamestown Company, although like its mother company it remained a subsidiary of the City of London Corporation. Lynch's account puts a new spin on the old tale of Puritan settlement in their "City on a Hill," the "New Jerusalem" of Puritan yearning, as their advocate William Bradford styled the Plymouth Colony experiment. He also makes clear some of the manipulations of investors and market forces that fueled the rise of the modern nation-states.

> In the opening chapter of our story as a nation, we read about native Indians, French trappers, Spanish conquistadors, sailors who sailed in the wrong direction, soldiers of fortune, explorers in coonskin caps, and Pilgrims at the first Thanksgiving dinner. But behind the scenes, somebody had to pay the bills for the ships, the food, and all the expenses for these adventures. Most of this money came out of the pockets of . . . investors. Without them, the colonies never would have gotten colonized.

In those days . . . if you wanted to go into business in the royal lands, which was most of the land on earth, you had to get a royal license, called a "Charter of Incorporation." These licenses were forerunners of the modern corporation, and business people couldn't operate without a charter or a piece of somebody else's charter. . . . The [men who] founded Jamestown had a charter. And once you had [that] you had to look for the financing. That's where the earliest stock market comes into play.[8]

Early in the seventeenth century, the Dutch East India Company hit on the stratagem of selling shares in public companies to fund their nascent multinational enterprises. The funds so raised were used to outfit and man ships bound east to buy a variety of goods, particularly spices and tea, both in great demand in Europe. The investments were solid: money was made in impressive amounts, leading the City of London Company and its incorporated subsidiaries to follow the Dutch example. Their interest was increased, of course, by the success of the United Dutch West India Company, which plied its trade in the Americas and was the corporation that funded an expedition to the lands of the Manhattan people. The Dutch adventurer Peter Minuit, who was employed by the Dutch West India Company as a kind of advance man, bought the island for a few glass beads, worth about twenty-four guilders, and laid claim to what would become New Amsterdam until the Dutch corporation gave up its holdings in the New World. As Lynch wryly comments, it was too bad that they didn't stay around long enough to reap the benefit of "owning all that downtown New York office space."[9] Even more ironic was the fact that Minuit conducted the transaction with a Susquahanna hunting party who hailed from territory in what is now known as Long Island. However, the Dutch West India Company felt it owned the whole nine by thirteen miles. The local people didn't object, of course, being oblivious to the transaction.[10]

Early market economies were chaotic. Investors were fickle, vulnerable to seduction, and willing to take risks. Often they handed out funds on the basis of recommendations that often as not were bogus. Lynch tells of one clever Englishman of that tumultuous era who convinced the French king that he could wipe out France's national debt and fill its coffers by floating national currency. His ploy worked so well that a bogus scheme he came up with later, raising large amounts to fund something he called the Mississippi Company, yielded him great wealth and netted a minus balance for the hopefuls who had bought his stock offerings.

A similar fate seemed in the offing for investors in the Virginia Company of Jamestown. The colony suffered reverses, some beyond anyone's control, some because those sent to secure the investment and make good on it refused to do so. From the beginning the settlers let improvements fall into severe disrepair and failed to raise crops sufficient to feed themselves—much less raising export crops as the Company expected them to do. Smith wrote extensively about the matter, being of the opinion that those sent to secure the colony and make it profitable were incompetent at best, and the laborers self-serving and uninterested in honest labor. Thomas Dale had been sent to whip the Jamestown settlers into shape, and whip he did. Under his dictatorial, tyrannical governance the enterprise seemed well on its way to profitability. However, management back at the City of London Company was beginning to balk at throwing more money into what seemed a dead loss. To renew investor enthusiasm and secure an infusion of much-needed cash, Dale set sail for England with Pocahontas, known in England as Lady Rebecca, and tobacco in tow. Sir Thomas Dale—acting governor (or "marshal") for the Virginia Colony—and his backers in England had quite a dog-and-pony show planned in which they would showcase the lady and, on Rolfe's part, peddle the new Virginia blend.

Of course the king wasn't happy about tobacco exports from the colonies, nor were most of the gentlemen of the City of London Company, an attitude that necessitated some agile lobbying by the Virginia Company of Jamestown contingent. Jamestown and the nearby homesteads needed some powerful backing, and one of their ploys was a public-relations blitz. The entry of Lady Rebecca—first

Christian, Anglicized Indian princess, a model of what they called "civilizing"—was choreographed to maximize public interest in the venture and secure the means to expand.

Disembarking at Plymouth, the party was met by such notables as Sir Lewis Stukely, the vice-admiral of Devon, along with a party of high officials and their lady wives. Perhaps Lady Rebecca looked in vain for Captain John, as some romantics claim, hoping he still lived, perhaps knowing as much. She would meet him in time, and at that meeting make clear her reasons for aiding him during his time in her country, and why she hoped to meet him on the dock.

As it was, she and her party were engulfed in a crowd of strangers, entirely at the mercy of those she had little reason to trust. One wonders if she recognized that she was a showpiece, a merchandisers' ploy. But even if she had—and she was intelligent enough to at least suspect that something like that was happening—as a *weroanskaa* it would have seemed only proper; as a medicine woman, ambassador, and spy it would have provided welcome access and cover. Rebecca, née Matoaka, called Pocahontas, "Mischief," entered history, where she would be known as the love-besotted Indian maiden Pocahontas.

The movements of the *manito aki* are ever intriguing: had it not been for the pressure of the bottom line, had the company's balance been in the black rather than in the red, had a less able public-relations manager planned their Virginia branch's next fund-raising ploy, Rebecca and her son would likely have remained back at Varina, the Rolfe homestead, or perhaps awaited Rolfe's return back in their home village. Ladies did not usually accompany their husbands on business trips at that time. But this lady was in herself a large part of the reason for the voyage. Sir Thomas Dale had often asserted his determination to spend the rest of his life in Virginia; whether with or without his English wife remains unclear. He had established Henrico and nearby holdings to that end, made certain that roads and bridges were well built to enable easy land access as well as access by river; he had even made overtures to Wahunsenacawh seeking a "marriage" with yet another of his "daughters" to secure the peace. Nevertheless, he returned to England to raise money to continue a venture that he would shortly abandon. Regardless of custom or aristocrats' wishes, when manito moves along in a certain way, anyone caught in its current is carried along.

In a very odd way, the modern concept of history and the traditional native concept of manito converge to make a certain meaning real in worldly time and space. Because of the life work of Pocahontas the meaning of their mating as economics and religion created a new nation, and a new world. At that level of function, Pocahontas was Amonute, a medicine woman whose secret being and greatest source of *powa* was *apook,* tobacco. Put another way, she was the carrier of the dream wheel, and what powered the dream was the sacred plant that grew from the head of First Woman and was given to human beings to ensure our connection to her knowledge.[11]

The twists and turns of the nascent colonial enterprise, generated by political necessity at the time, became a template for similar enterprises for centuries to come. The historical record is clear: the adventurers had often falsified or put a particular spin on their reports— customarily titled *True Relations.* They had disposed of documents that might contradict these carefully constructed official reports as well. It was business as usual, and their methods were not unlike those employed by modern corporations. But there were differences, chief among them the existence of a powerful monarchy. At that time the monarchy had a great deal more control of Parliament and of fiscal matters than it presently possesses. The Virginia Company was what today might be called "a wholly owned subsidiary" of the City of London Company, itself a massive guild system that largely controlled enterprise throughout the kingdom, particularly in the City and in its international transactions. It was precisely that monarchal power that led to the revolt that took place under Cromwell's leadership in the mid–seventeenth century, and led to the beheading of Charles I.

Self-determination and private property were concepts whose time had almost come. By 1619 land was deeded, by order of the king, to planters who had holdings. But at the time that Pocahontas was in England the maneuvering between monarch and emergent capitalists was still well below the boiling point. Not only was the idea of personal ownership of property by classes other than aristocracy a while off, but the idea that business concerns could be privately held was beyond imagining. The City of London Company, which was a conglomerate of the City's various guilds, functioning since medieval times, was the holding company for all business enterprises emanating from London. Privately owned corporations such as exist in the modern world were yet to come, although the very recent first stock market in Amsterdam foreshadowed what would in centuries to follow become an ordinary fact of modern life. No one involved, other than perhaps Rolfe and his lady, had any idea of owning their own land or business in that year of 1616 when the Jamestown representatives of the Virginia Company hit town.

Sir Dale, for example, had made all the improvements to the local infrastructure, and built himself a home, a parsonage, and the first Indian school, but he saw the settlement as being owned by England. This meant that it was worked by the indentured English men and women sent over for that purpose. All goods were the property of the company and were marked for export, all proceeds to return to the Virginia Company of Jamestown. And although at a practical level the whole was the concern of the City of London Company, that enterprise and any land it settled or used in any way was the property of the monarch. The royal charter granted rights of use, not ownership. This may be an alien concept to contemporary Americans, but it is a system that still marks much of the Western world.

Varina, the Rolfe holding, had been a wedding gift to Pocahontas from the Powhatans, perhaps from Wahunsenacawh himself, more probably from the *matchacómoco*. This meant that the land was not part of that governed by the rules of the royal charter, but that point might have been moot in Rolfe's mind at the time, and certainly became moot in the course of the next couple of decades, although, given the king's attitude toward Thomas Rolfe, Varina, a gift to Pocahontas from Wahunsenacawh, would pass to her son regardless of James's preferences in the matter because it was Indian land.

THE POPULAR LOTTERY

Even though the settlement and the great river on which it was set bore his name, James I was uninterested in investing cash from royal or personal coffers for its expansion and maintenance. The thousands of pounds of tobacco the men of Jamestown has sent or brought to England and the money derived from its sale did little to persuade him. After all, he hated tobacco. Given the new economic forces taking hold in Europe, the major investors in the Virginia Company of Jamestown formed a coalition of merchant and clergy to raise money. They called this group the Popular Party, and set about the task of securing funds. One goal was to collect money promised but not yet paid by stockholders. In some cases the pledges amounted to large sums, and the venture was in need of every shilling. They also encouraged new investors to pony up.

One of their more lasting fund-raising schemes was to host a lottery to be held at St. Paul's Cathedral (the old one rather than the one now standing). This was in 1615, and the stated purpose was to use the proceeds from the event to make a success of the Virginia Company Jamestown "as a worthy Christian enterprise."[12] One of the major chroniclers of Jamestown's earliest history, Samuel Purchas, rector of St. Martin's Church in London, described the lottery event thus:

> The great standing lottery was draine [drawn] in 1615 in the West end of Paul's churchyard . . . in which the Prizes were proportioned from two crowns (which was the least) to divers thousand . . . and paid in money or in Plate there set forth in view, provided that if any chose money rather than Plate or goods for paiment in summes above ten crownes, he was to abate the tenth part. The orders of this lottery were published, and courses taken to prevent fraud. [13]

The gentlemen of the Popular Party were so pleased with the success of this event that in 1616 they determined to embark on an even more elaborate fund-raising program. This idea had been floating around in clerical circles for some time: it was to bring Princess Matoaka, Lady Rebecca (as they formally identified her), to England as a showpiece of the success of their Indian catechetical program and therefore of their determination to place the bit in the despised Papist—Spanish—mouth, as they so colorfully insisted. Their plan was to bring the lady over and, by parading her before royalty, clergy, and merchants, attract even more money and colonists to Virginia. In prime seventeenth-century public-relations style, they made arrangements for her to sit for a formal portrait, be received at court, and be entertained, along with ladies and gentlemen of substance, at a masque to be written and produced by the popular poet and writer Ben Jonson. Thus she would amuse the rich and famous of the day, promoting their enterprise and giving greater glory to God and country. The curtain on the final act of the Pocahontas masque went up June 12, 1616, when she and her husband, son, and party of Indians disembarked. They made their way to London by carriage, and were settled at an inn named La Belle Sauvage, located on Ludgate Hill, near St. Paul's Cathedral.

Thus began the next stage in the company-sponsored public-relations campaign. The campaign specified that their own *belle sauvage* be presented to three major branches of society: the government, represented by the king, the royal family, and the court; the church, represented by the most notable among the divines; and the trades, represented by the country's most prominent merchant families. One supposes that among those she received in London were Popular Party sponsors such as the bishop of London, John King; poet and divine John Donne; and the Earl of Southampton, who was a friend and patron of William Shakespeare.

The gentlemen of the Popular Party asked John Smith to approach Queen Anne on the matter of her receiving Lady Rebecca at court. While many of them were out of favor, Smith had dedicated one of his works to the queen; it was felt that the dedication, along with his personal familiarity with Pocahontas, had the best chance of securing royal notice. His letter was effective; the queen appointed Lord and Lady De la Warr (Lord De la Warr was then formally the governor of Virginia Colony) to serve as the royal visitor's sponsors, attend her at court, and escort her during her time in London, introducing her to the proper people and taking her to the theater and other occasions.

She would be seen attending performances of *Twelfth Night* and *The Tempest,* two spectacles that must have seemed very familiar. All was orchestrated to make her the toast of London, and by all accounts it worked: Lady Rebecca had a good command of English, dressed attractively, comported herself with quiet dignity, and danced gracefully. The gossips were largely of the opinion that this was all quite marvelous; they said that she was indeed "a comely young woman—although of dusky hue," and was altogether worthy of the time and money the Virginia Company and its divines, Alexander Whitaker and Richard Buke, had lavished upon her.

In the end, of course, the result of the queen's effort was that Pocahontas remained steadfastly pro-English. Conscious of her stature throughout her life, she seems to have concluded that her loyalty was required as a matter of honor. She was treated with the respect that befitted a royal ambassador of a great nation to the court of another. So pleased with her reception in the capital was she, in fact, that she didn't want to return to Varina and the hard life of tobacco farming, the harsh disciplines of the rural church.

"The Virginian woman Pocahontas wᵗʰ her father Counſaillor [Uttamatamakin] haue ben wᵗʰ the King and graciously vſed, and both ſhe and her aſiſtant well placed at the maſke, ſhe is vpen her return (though ſore againſt her will) yf the wind wold come about to ſend them away," one courtier confided to another a few days after the event.[14]

A MEETING OF MAGICIANS

Meanwhile, Lady Rebecca, Powhatan ambassador and intelligence agent for the manito, was engaged in her own mission. Soon after her arrival, she conferred with Sir Walter Ralegh, who escorted her to the Tower of London for a visit with the famed wizard the Earl of

Northumberland. Pocahontas had met the earl's brother, George Percy, when he was part of the company; Percy himself also called at the Rolfe rooms at the Belle Sauvage inn. Why Ralegh paid court to Lady Rebecca raises questions. Himself a spy for Elizabeth I, a member of the Enochian circle of the queen's astrologer and wizard Dr. John Dee, Ralegh likely wanted information about Algonquin arcane knowledge as much as the Natives desired it from English wizards like Ralegh. While he was out of favor with the court, his network within and outside of court circles was extensive and far-flung. Given Ralegh's lifetime of involvement with magical and alchemical studies, as well as with Indians brought to England—which is how tobacco use was first introduced there—there is every likelihood that he was in possession of information about the supernatural entities and powers that graced the Americas. He not only had crossed over himself, but had been the driving force behind England's essay into virgin land. Perhaps he was hopeful of getting some information from the Indian ambassador's party about the lost Roanoke colony, which he had founded. Then again, maybe he was even more interested in gathering some information for his own network; maybe he hoped to gain occult information from the *manito aki*. Sir Walter was a strong advocate of the widespread use of tobacco, a plant that he knew had sacred properties; he himself was a constant user, and had long set fashion in smoking implements such as pipes, snuff forks and holders, and other paraphernalia that upscale tobacco aficionados of the era affected.[15]

For her part, Pocahontas may have been curious about the famed adventurer; and as a priestess of a high degree of spiritual attainment, she is likely to have wanted to exchange notes about their respective hidden traditions. Given Ralegh's fame, it is likely that Rolfe, a fellow addict, alchemist, and adventurer, would have told her about him. His name would have surfaced naturally enough because of his connection to the lost colony, Roanoke. It must have been of great interest to the *manitowinini* to obtain high-quality information concerning the magical power that the English wielded. They had seen for themselves how great this power was, in the tools and ships, the clothing and adornments, the metals and glass-blowing marvels these strangers from across the water carried or produced. The Council of Elders instructed Uttamatamakin to be sure to meet both the English king and the English god the English had told them about. These were pagan people, free of the concept of a single monarch or single god ruling over all. In a variety of ways, their idea of the divine, the supernatural, was that it was light-years closer to the human realm than it was conceived among English theorists or divines in the seventeenth century. In Algonquin experience the divine was more directly and manifestly involved in all human life and the lives of other entities. The Algonquin world was what ethnographers have styled "animist," which means that they held that all the world's creatures were sentient, intelligent, and not necessarily benevolent when it came to the affairs of human beings. Surely the most pressing question for them was the identities and inclinations of the spirits who endowed the aliens with the power to bring about the "ruin and devastation" of the *tsenacommacah*. Considered in terms of another discourse, they were far more connected to and interested in the workings of the implicate order than connected to and interested in the explicate. In a sense, their universe was more Shakespearean than ours.

Given this, Pocahontas and her companions would have observed England as they observed their homeland. They would have looked for the manito, how manito presence and powers were manifested—which included society's rules and customs, architecture, city layout, food and drink. They would have closely examined as well the greater environment within which humans and other intelligences operated, taking note of how the otherworldly beings made known their desires, their dislikes, their inclinations, and their preferred manner of contacting the human realm.

Interviewing Englishmen who were rumored to be the most adept among their priests would have been high on the list of tasks the Powhatans, especially Rebecca, Uttamatamakin, and Matachanna, had been given by the *matchacómoco*. As they had been questioned several times about the whereabouts of the lost colonists of Roanoke, the name of Ralegh would have come up. Similarly, attendance at social functions, particularly the evening at court, but also at the theater, would have been of particular interest, because these were, in their experience, the most sacred events. At Algonquin ceremonial occasions, various manito were courted, fed, evoked, exhorted, and given generous gifts of tobacco so they would be drawn toward human circles and kindly disposed to them. To a great extent the calling of manito to consort within the human community, to be its guests, is the best definition of ceremonial or sacred events. The gathering at the court of King James, at which Jonson's masque, along with dancing, and lengthy feasting were featured, was the pinnacle, ceremonially speaking. For the visitors from the *tsenacommacah,* this event would have been analogous to gatherings at the *quioccosan* in Werowocomoco, their own seat of royalty, which to them meant the seat of the sacred—that is, the seat of manito presence.

Princess Matoaka, Lady Rebecca (as she was formally identified in England during her tour), found these manifestations and presences enchanting. Not so Uttamatamakin. He loathed the place and all the beings in it. To him the manito of the English land was like Caliban in Shakespeare's recent drama *The Tempest.* To the royal Powhatan ambassador, however, it was more like Ariel, the water-spirit power both visible and invisible who helped Prospero, the magician. One wonders if she took note that he, Ariel, was enslaved and compelled to do Prospero's bidding in order to gain his freedom. One further wonders whether she remarked on the similarity of Prospero's dear daughter Miranda to herself. She must have known that the English thought she was Powhatan's daughter, which in a sense was accurate enough, spiritually speaking. The Powhatan used the English words *father* and *daughter* to accord, as they thought, with English practice, but the Powhatan meaning differed wildly. Patriarchal dominance, an established principle in Europe, was unthinkable in Indian Country. The Powhatans may have believed that bi-gender balance was common in England because it was so among them, or they may have concluded that in this as in most other matters the English were, well, strange.

As the ethnohistorian and anthropologist Frederic Gleach points out, to the people of the *tsenacommacah* the English word *father* meant "leader," *mamanantowick,* central greatest chief, *kitchee weroance.* That certainly described the role of Wahunsenacawh. "Daughter" designated a child, less powerful because of age, as well as a female child ("daughter") rather than a male child ("son."). That described Pocahontas in her relation to Wahunsenacawh accurately. Thus James I was "father" of his people. Those who were women were his "daughters." "Mother" would have meant clan mother; the members of that clan would have been her "children." The southern Algonquins from the area where Roanoke was established referred to James I's predecessor, Elizabeth I, as *weroanskaa,* or "greatest chief" (feminine).[16]

The Powhatan ambassadors to England must have believed James I to be a great, great magician, a *mamanantowick,* possessor of unbelievable *powa.* Nothing they beheld or heard would have indicated otherwise. Wahunsenacawh, who held the title Powhatan at the time of Pocahontas, was considered the principal chief or head of the Powhatan Alliance. While the office of head of sacred state passed down matrilineally, there were those, such as Wahunsenacawh, who possessed additional powers. In the case of the man known as Powhatan to the English, William Strachey mentions the "great fear and adoration" with which the people carried out Wahunsenacawh's directions.[17] "This seemingly absolute authority was probably partly due to Powhatan's holding shamanic or priestly power in addition to his inherited position."[18] Strachey tells us that the *mamanantowick* projected an "impression of divine nature . . . in terms of an infusion of godliness; this state of [the *mamanantowick*] was obvious to the English as well as to the Indians." Gleach cites Irving A. Hallowell, who observed how the Ojibwaj (Western Algonquin) saw such a person: "Men who are believed to have acquired much power from other-than-human sources occupy a special position in Ojibwa culture. Since they have both the power to cure, as well as to kill [by supernatural rather than physical means], people's attitudes towards them are ambivalent: they may exercise a great deal of personal influence because they are feared, and may outrank other men in power though not [necessarily] in material wealth or social ranking."[19] Considering that James I was the King of England, Wales, and Ireland, possessed of wealth and standing of staggering proportions, Pocahontas would have believed him to be a *mamanantowick,* of far greater power than any the *tsenacommacah* could have imagined. Small wonder she expressed her deep desire to remain in England. Far from being a "sellout," as modern Nativist radicals have accused, she was intent on her mastery of her craft and deepening and expanding the Indians' relation to the unseen powers that be. In her understanding, politics was manito centered and influenced, if not directed, and economics were an extension of *powa.* She, Uttamatamakin, and the others of the Powhatan delegation must have believed that James I was a stupendous example of the kind of ruler familiar to them, as such a one was described by the French scholar Marc Lescarbot. In 1897 Lescarbot commented on the merging of the powers of church and state among the Algonquin of French Canada: "It happens sometimes that the same person is both *Autmoin* [which, like *mamanantowick,* means a combination of priest, prophet, medicine person, soothsayer, dreamwalker, and close ally of powerful manito] and *Sagamore* [head of state, similar to *weroance*], and then he is greatly dreaded."[20]

AN ENGLISH MASQUE

Lady Rebecca's evening at court was carefully orchestrated. Though it would take place at the end of the tour, early in her visit the company made arrangements to secure her reception at court. Should they succeed in that, their investment in bringing over Lady Rebecca and her party and providing quarters, meals, attire, and all the associated expenses was going to pay handsomely. It would go far in securing investment capital from James I, or so they hoped, and the financial backing of the royal establishment. They had designed their three-pronged strategy to ensure that each segment of affluent England would reinforce the other in making their offering seem a wise investment indeed.

To meet this objective, they called on John Smith to write to the queen and request that Her Royal Highness invite the Indian princess to court. In his letter, Smith makes much of how the Indian child had intervened to save his life, and the lives of the others who comprised the Jamestown community at various times. He assured Her Highness that "Pocahuntas" was a "tender virgin" who had intervened on the behalf of the king's subjects at great peril to her own life.[21]

He put every bit of his considerable persuasive skill into the letter, and it yielded results: Lady Rebecca Rolfe, Princess Matoaka, escorted by her "father-councilor" or "assistant," Uttamatamakin, was invited to the palace for a midwinter gala that would include dinner and dancing and would feature a masque, *The Vision of Delight,* written by Ben Jonson, Queen Anne's favorite playwright. John Rolfe was restricted to the outer fringes of the audience; he was seated in the gallery during the evening, largely because, to the king's pique, Rolfe, a commoner, had presumed to marry a woman of royal blood, even though that blood was Indian. No fan of interracial marriage, James frowned on the alliance on those grounds, of course, but breach of class restrictions was the major focus of his anger. Thus the Princess Royal, accompanied by Uttamatamakin, was seated at the high table next to the queen. Appropriately gowned and coifed for the occasion, she danced gracefully, conversed charmingly, and by all accounts comported herself as suited a royal princess, however "salvage" her origins.

Although Smith's plea was predictably self-aggrandizing and self-serving, dramatically portraying Wahunsenacawh's "fury" and "salvagery" and his own stalwart dignity, he is perceptive, and prescient, when he writes,

> The most and least I can do, is to tell you this, because none so oft hath tried it as myself, and the rather [Pocahontas] being of so great a spirit, however her stature [status]: if she should not be well received, seeing this Kingdom [England] may rightly have a Kingdom [Virginia] by her means; her present love to us and Christianity might turn to such scorn and fury, as to divert all this good to the worst of evil.[22]

Pocahontas's reception at Plymouth and in London caught the attention of the London smart set, as the company's public-relations campaign intended. A celebrity as well as a curiosity, she was visited by dignitaries, artists, and courtiers. Given that they well saw the differences between themselves and Rebecca, the savage princess however royal, and doubtless comported themselves to make their views obvious, these visits must have put her under great stress. When one of the most celebrated among them, the famed poet and playwright Ben Jonson, came to call, he chatted with the royal ambassador for a brief time, then fell to gazing at her without speaking. After a half hour or so of this, she left the room, leaving him to his musings and bottle of wine. The strain of constant entertaining, the culture shock, the tension of maintaining her focus despite her several roles, and the filthy city of London, where air, water, streets, buildings, animals, and human beings of every class went unwashed year after year, began to show on her. She was forced by repeated illness to take to her bed, and her overall health began to fail.

Pocahontas came to England from a nation that bathed daily, so she and her companions must have suffered greatly from their own unwashed condition, exacerbated by the heavy clothing that went unwashed and unaired; it was deep winter in England. The danger of the pervasive presence of decaying animal and vegetable matter that was simply thrown into the river and local sewage streams, human and animal waste that lay on the walkways and filled the muddy streets, and the heavy pall of smoke from thousands of charcoal hearth fires combined with the dense fog of a London winter surely added to the visitors' misery and contributed to her worsening condition. Various efforts were made to improve Pocahontas's condition; while the gentlemen of the Virginia Company were concerned for her health, it was also impossible for them to arrange her return to Virginia, and end their cash outlay for her expenses, so long as her illness prevailed. Arrangements were made for her to spend time on a large country estate in Brentford, nine miles west of London. Syon House, as the estate was named, bordered the Thames for several miles. Rebecca must have felt almost at home amid the beautiful buildings, manicured rolling lawns, large stands of trees, and extensive gardens. It was the queen's custom to spend the warmest summer months at Syon House, one of the great estates of Lord Northumberland, with her retinue of ladies-in-waiting, Lady De la Warr no doubt included in attendance.

There the Powhatan medicine woman, Pocahontas, may have met the Green Knight, perhaps in his summer guise as Puck, the goatlike god of the forests. Perhaps she found fairy circles in the grass of some clearing in the trees, or heard the laughter and singing of the fair folk echoing on the air from their unseen feast and dance party. She may have run into the spirit of King Henry VIII, whose corpse had lain at Syon House in its great hall awaiting the ship that would carry the king's remains upriver to the royal final resting place. This occurred in full summer, and the heat during the time that the body lay bloating caused a horrifying moment: the king's corpulent corpse exploded, spewing royal parts in a dreadful stench over the room. The ghosts of Anne Boleyn and the myriad others who died because of Henry's determination to sire a male child who would succeed him must have felt grimly amused at the sight. As such highly charged events make a noticeable impression on what Yeats called the *spiritus mundi,* if far sight was among her gifts, this Indian *powatanep,* medicine woman, would have learned quite a bit of English history firsthand. Perhaps this was another reason she longed to remain in England longer. Perhaps she and her circle held the Powhatan equivalent of séances and watched the English past flow by as we watch television or films. Perhaps they cast enchantments, prayed, evoked, made offerings, and in general worked their magics on the local spirit people, the local *powa*-fields, the local *manito aki,* Faerie.

One of the more significant events occurring during Lady Rebecca's stay in Brentford was John Smith's visit. He had studiously avoided seeing her, although by his own account many of his friends had gone to her rooms in London. A few days before he was to set sail somewhere, he overcame his reluctance; although he makes no mention of it, his avoidance might have been on account of the lies he had told his "father" Wahunsenacawh, his savior Pocahontas, and other "good friends" back at the *tsenacommacah*. So it was that, however tardily, John Smith turned up at Syon House, perhaps fearing that if did he not, somehow the queen would discover his lapse and discount whatever credibility he had gained with Her Royal Highness.

> Being about this time preparing to set sail for New-England, I could not stay to do her that service I desired, and she well deserved. But hearing she was at Branford [Brentford] with divers of my friends, I went to see her.
>
> After a modest salutation, without any word she turned about, obscured her face, as not seeming well contented. And in that humor, her husband with divers others we all left her two or three hours, repenting myself to have writ she could speak English [to the queen].
>
> But not long after, she began to talk and rememb'red me well what courtesies she had done, saying, "You did promise Powhatan what was yours should be his, and he the like to you. You called him father, being in his land a stranger, and by the same reason so must I do you;" which though I would have excused I durst not allow of that title because she was a king's daughter. With a well set countenance she said, "Were you not afraid to come into my father's country, and cause fear in him and all his people? And fear you here I should call you father? I tell you then I will, and you shall, call me child, and so I will be forever and ever your countryman. They did tell us always you were dead, and I knew no other till I came to Plymouth. Yet Powhatan did command Uttamatomakkin to seek you, and know the truth, because your country men will lie much."
>
> . . . Divers courtiers and others my acquaintances hath gone with me to see her, that generally concluded they did think God had a great hand in her conversion, and they have seen many English ladies worse favored, proportioned, and behaviored.[23]

There is a general consensus among writers and scholars that Pocahontas treated Smith so discourteously because she was shocked that the man she had loved so truly could have been such a cad. She felt betrayed, this school of thought goes, a woman scorned. So first she refused to speak to him; then, having punished him with her silence—in front of his friends and her husband—for a few tiresome hours, she called him aside and bitterly accused him.

This analysis, however, seems to ignore the report Smith himself wrote. There is nothing in Pocahontas's remarks that refers, however tangentially, to romantic love. She is clearly upbraiding Smith for dereliction of duty, for dealing dishonorably with Powhatan, the Great King and his father, to whom he had sworn loyalty. She is aware of the many instances of his duplicity, and rightly confronts him with her knowledge and her shame that he could act so.

She is not ashamed of herself, but in the Indian way is embarrassed for *his* crassness, ashamed that he should behave so before witnesses. She makes the nature of their relationship clear: "'You did promise Powhatan what was yours should be his, and he the like to you. You called him father, being in his land a stranger, and by the same reason so must I do you;'" she tells him, making it clear that Smith's relationship to Wahunsenacawh was as son to father, and because Smith had accepted this characterization of the relationship, he accepted the law of kinship: that what was his would be the Powhatans'. This, she says, is the basis of their relationship: she must call Smith "father," being a foreigner in England. That being the case, he, Smith, owed to her the same courtesies extended him by the Powhatans, and by the Great King himself: to greet her, to escort her, make her relationship to him known publicly—as his daughter—and therefore to make certain that she would be as much an Englishwoman as any woman born there. That he did not, that he failed to meet her at Plymouth and then failed to greet her immediately upon her reaching London—at the very least, these were the crimes against the fundamental law of relationship and the consequent obligations relationship, kinship, demand. He failed abjectly, and, she finishes triumphantly, the Powhatans know how the English lie, and have sent Uttamatamakin—and her royal self, though she withholds that bit of intelligence—to locate Smith and ascertain the facts, given that they are convinced by their experience that duplicity is an English cultural characteristic. Her attack couldn't have been more scathing. That anyone, either Smith or later writers, could mistake her accusations as words of anger felt by a woman scorned is one of the puzzles of the Pocahontas story in American cultural lore.

This is the heart of the matter of the Pocahontas legend and the truth of "Pocahontas"—an Algonquin woman named Matoaka, then renamed Rebecca. However tale tellers want the stranger to be the same culturally as the teller, in matters of different sets of cultural consciousness, it is a highly unlikely occurrence. The tendency to make assumptions of sameness is probably the single most telling indicator of the disaster that was to come from the collisions of two worldviews. The one that triumphed, at least seemingly so in the short term, seems no more guilty of making that unsafe assumption than the one that fell into ruin.

Smith himself seems to have been oblivious of the deeper significance of his *tsenacommacah* friend and ally's remarks, for he blithely discounts them. He could hardly view her as his close relative, he believes, because she was royalty and he was not. But then, our social conditioning is so great, however adventurous we think ourselves to be, that his very obliviousness speaks volumes about the truth of that tragic chapter in Algonquin history.

After some weeks in Brentford, Rebecca seemed recovered sufficiently for the party to begin making plans to return to Jamestown. Although it wasn't in the company's original plan, someone seems to have determined that the young couple should visit the ancestral Rolfe estate. The visit to eastern England, an arduous journey up from London, was difficult. Making it may have exacerbated Lady Rebecca's condition, and little was made of the visit by the Heacham Rolfes or the London gossip circuit. While they were there, however, a portrait of Lady Rebecca and her son, Thomas, was made. There was one other portrait of her painted, a formal likeness in which she is costumed in high Puritan fashion. Perhaps this is the outfit Lady Rebecca wore to the masque at Whitehall Palace, an event that occurred in December, after the Rolfes returned from Heacham.

Whether this is so depends on the evening fashions of the period. If stiffness was the current standard of ladyhood, then, so garbed, Pocahontas would have been m'lady indeed. Two versions of the formal portrait have survived the centuries—one, unsigned, in oils; the other, an engraving based on the oil, signed on the lower left-hand corner, SI. PASS: SCULPT. The signature informs us that the piece was done by a young Dutch engraver, Simon Van de Passe, who was the designer of the work. The name of the printer who issued the small edition, Compton, is engraved together with Van de Passe's name. There is some discussion of whether the painting was based on the engraving. If so, it was done in 1616, the date the engraving was made. An oval border surrounds the figure. It bears the title: MA-TOAKA ALS REBECCA FILIA POTENTISS PRINC : POWHATANI IMP : VIRGINIÆ: "Matoaka, alias Rebecca, Daughter of the Great Prince, Powhatan, Emperor of Virginia." The figure is dressed in a heavily starched gown that features a high, starched collar edged with pointed scallops. At the back of her head, the tip of the collar reaches the brim of her small, high crowned hat of beaver. The hatband is wide, and the brim narrow. On the left side rises a brush of hair—beaver, perhaps. In one hand a three-plumed white feather fan is propped gracefully, signifying her office as Beloved Woman—a significance perhaps known only to her and the Powhatan members of her party. That she holds this staff of office, as it were, for this formal portrait says much of her self-image and the image she intended to transmit to the English rulers and high society. Her expression is serious but gentle, her hair pulled back, revealing earrings dangling from each lobe. They may be those gifted her by the "Wizard Lord," as Northumberland was known, when, accompanied by Ralegh, she visited him in his apartment in the Tower of London. Just below the figure where a narrow light band divides it from the bottom of the oval is inscribed: "Ætatis suœ 21. Å." and beneath that, the year, 1616, informing viewers that the engraving was made on August 21, 1616.

Their visit to Heacham completed, the Rolfes and their party set off once again for London, where they had been invited to attend the queen's favorite masque, the Twelfth Night winter holiday masque, *Vision of Delight*. Elaborately coifed and gowned, Lady Rebecca was seated among the other royalty as befitted her rank as daughter of the high prince, emperor of Virginia, Powhatan, and as royal ambassador to the court of James I. The Lady Princess was well enough received, having made the acquaintance of the queen, who was the true host of the festivities. She was presented to the king, Anne of Denmark's husband, and enjoyed the feast, dancing, and masque. She may have been keeping extensive notes—in the superbly trained memory that those raised in Oral Traditions develop. Mentally comparing this performance with those her people held back home in the *tsenacommacah*, Pocahontas would have paid careful attention to any details that seemed to pay reference to or contain particularly revealing instances of *powa* and/or manito presence. It being the Christmas holiday season, one of the highest holy seasons of the Christian year, she would have been doubly alert to nuances and resonances of the sacred. At her side, Uttamatamakin would have been similarly occupied, and they may have made soft remarks, or exchanged looks or small gestures, as one or the other recognized some gesture or tone. Surely their discussion after the event, when they shared whatever information they thought appropriate with the other members of their party involved in the espionage, would have been a perfect opportunity to share their gleanings and to plan further strategy.

ONE SPY'S POISON IS ANOTHER SPY'S DELIGHT

While it is likely that Uttamatamakin was little impressed with what he learned, Lady Rebecca may have been. It seemed clear to those in attendance, or afterward visiting her, that she was loath to leave England. For whatever court gossip is worth, the idea that she wanted to stay became one of the accepted parts of the story; to later generations of English and Americans her wish to remain there demonstrated how unique she was among Indians, how civilized, how white. However, whether she would or would not, the decision was not hers to make. The London Company and the investment group directly responsible had grown weary of the money they were lavishing on this visiting model of English efforts to civilize the savages. They wanted the party returned on the strongest west wind so they could be quit of the financial drain and move on to improving the bottom line. Rolfe himself, along with his cronies Hamor and Argall, were anxious to return as well, no doubt eager to take up their new positions and parlay them into real wealth and status.

Meanwhile, the lady's health was once again in decline. Whether it was the English winter or the demands of the London social season, or whether there were subtler forces afoot, is not clear. There are no records of symptoms of her illness other than that it was debilitating and mysterious; in the absence of specifics, different diagnoses have been offered over the centuries. Smallpox was frequently suggested as the cause of her death, but that seems unlikely. Respiratory illness, perhaps a devastating case of the ague and fever that had come from the Continent into England and reached epidemic proportions during the last months of her stay, has also been suggested. The most diligent researchers agree that the cause of her death is unknown; records indicate only that she was very ill by the time the party went aboard their ship for the return trip.

There is another possibility that must be explored: she and Matachanna, and even Thomas, her son, may have been poisoned. As a method of murder it was not unknown in those times; it had achieved some renown, being the weapon of choice by such celebrities as Lucrezia Borgia, the Italian aristocrat. Popular in court circles and common ones alike, it had the benefit of being somewhat controllable. A judicious poisoner could, given the knowledge possessed by most practicing alchemists of the time, administer doses so as to cause slow but eventually lethal decline. As in all cases of murder, method, motive, and opportunity are the key elements, along with a corpus delicti. The last named, a corpse, was indeed in evidence, although hastily enough dispensed with. That Thomas survived seems to have been mostly because Argall and Hamor prevailed on Rolfe to leave him in the care of Sir Robert Stuckley, who by all accounts was a vicious man. It is said that taking the child may have been the only kind act Stuckley ever performed, but when one ponders the intricacies of international diplomacy and the plots and counterplots inevitably involved, the murky waters begin to clear.

Jack D. Forbes, Abenaki (American Indian) Professor of Native American Studies at the University of California at Davis, shared his opinion that Pocahontas had been killed because she had a great deal of information about the monarch, which she planned to pass on to the Powhatans. He surmises that the murder took place aboard ship at the time Pocahontas's party waited to rendezvous at Gravesend. But the descriptions of her condition—"in decline," "a wasting disease," "mysterious disease," "failing health"—are suggestive. Equally so is the absence of medical reports, although some highly able physicians should have been summoned to treat her, as was proper for a person of her status.

That she rallied for a time when she was under the care of the lord and lady at Brentford, consorting with the queen's ladies, recovering sufficiently for the company to begin making plans for a return expedition to Virginia, tends to raise questions. So does the puzzling visit to the all but inaccessible Rolfe ancestral estate—to which Rolfe's claim is suspect. It seems that her condition continued in its downward spiral during the fall of 1616, when she was once again engaged in the social whirl that accompanied the fall return of the socialites and the court to London. There is no mention of her health at the Whitehall masque, or none that has survived. Either John Chamberlain, the irrepressible gossip, made no mention of it in his voluminous correspondence to his circle of courtiers, or such letters as contain mention have been lost.

The return expedition was mounted with haste and determination, and was ready by the first of February. Much to the chagrin of the Popular Party, who saw their reserves draining, the outward voyage was postponed a number of times, delayed because of the high winds that prevailed for three months. The party left London as soon as possible, although their hasty departure meant that they would be held up in Gravesend so they would arrive in Portsmouth at the same time the *Treasurer* did. It was during this interval at Gravesend that Pocahontas died.

The expedition that took the newly promoted company officers on to their new duties, and the colony—or company branch—was under the direction of their new deputy governor, Argall, who took his orders, at least nominally, from Lord De la Warr. De la Warr, of course, along with his wife, had been commissioned by the queen to look after the Indian princess-ambassador. Lord De la Warr was the governor of Virginia Colony. Men such as Sir Thomas Gates and Sir Thomas Dale were the acting governors; on-site managers doing the actual management would have been unthinkable for men of the high status of De la Warr. The true governor would, of course, have come from ranks high above; only a lord could hold such a lofty position in class-bound England. To him answered Rolfe, Argall, Hamor, and the entire colony. The point of the colony was primarily financial, although it had a strong political component that revolved around the power struggle in progress in England between the Puritan and the Anglican factions for control of the Church of England.

Any one of these men may have realized that the Powhatan contingent was involved in something other than innocent, primitive pursuits. The celebrity of Pocahontas, Rolfe's wife, with the intelligentsia, the wealthy and powerful, royalty, and the wizard Northumberland—a popularity that gave her and at least Matachanna access to ladies of every rank—may have caused alarm. While women in Elizabethan and early Jacobean English society were not among the movers and shakers, no one was in doubt as to their formidable prowess. Pillow talk then as now had its influence, and among those whose marital alliances were as much political as economic, and not at all romantic, the threat posed by her consorting with queens and ladies was no small one.

There was no lack of candidates for the position of murderer, or murderers. If Pocahontas was murdered, it seems more than likely to have been the work of more than one person—a conspiracy, if you will—and among the most likely candidates was John Rolfe himself. It was his decision, or so the story goes, whether he and his family would make the return trip despite his "wyff's" poor health. Surely, had he feared for her life, he might have appealed to Lord Northumberland, should petitions to De la Warr fall on deaf ears. He might have sold off his Virginia holdings to Argall or Hamor in order to raise the money necessary to support them in England. He had a brother living in London, the one who took over the rearing of Thomas, with whom they could lodge.

But it seems the good husband was determined to load his dying wife aboard ship, and wait out the few days until her life ended. Rather than distraught, Rolfe may have been relieved, and, not wishing one of the Powhatans—Uttamatamakin, for example—to realize he was far from grieving, kept away from her deathbed. Matachanna was herself gravely ill, and Thomas hardly better. Matachanna did not recover well enough to tend to Thomas, or so it is said, requiring Rolfe to desert his son at Plymouth, leaving him in the hands of a well-known scoundrel, at Argall's urging.

Whatever the reasons for the decision, financial, political, or criminal, when the high winds that had thwarted their attempts to set forth in February finally turned in their favor, midmonth in March 1617, Rebecca, Matachanna, and the child were carried aboard the *George,* flagship of the expedition. Accompanied by the *Treasurer,* now under the command of newly promoted Vice-Admiral Hamor, they made their way down the Thames to Gravesend. Happily, the party was not required to return on the *Treasurer,* the ship on which they had voyaged to England. With March and the drop in the winds, spring had broken in southern England, and everywhere flowers were breaking into bloom.

Dying, Rebecca must have seen little of the loveliness of her last spring. Her concern must have been for her little boy, and for the intelligence that only she could take back to the clan matrons, the *weroanskaas,* the medicine women and diviners, to Granny Squannit herself. She must have fretted in her final hours because she would be unable to complete her mission—at least in the flesh. Perhaps she hoped that Matachanna would survive to pass along the intelligence; perhaps she hoped she had shared everything needful to their charge with her sister-priestess. She would have known that the indefatigable Uttamatamakin would reach home and the male information he had collected would find its way to the right ears. That may have been her comfort. I wonder if she knew she had been murdered. Probably. If so, she would perhaps have laid a curse on the murderer, or have expected her relatives to take the necessary steps to restore the balance. Maybe she entered her girlhood Dream-Vision in her final moments, and walked off into the English countryside to seek the local spirits and supernaturals. She died on the spring equinox of that year. The supernaturals of summer had arrived, the powers of winter having returned to their place of rest.

By the time they reached the small village of Gravesend, mainly a garrison armed and manned against attack from Spain, it was clear that she wouldn't last much longer. Uttamatamakin carried her to the village, where it was hoped she might receive some medical attention—leeches, herbs, some physician's care. There was none to be had in this outpost, and at any rate she was all but dead by the time they reached the garrison. John Rolfe, said to be too distraught to attend his dying wife, waited as others tried in vain to revive her. Shortly before she died, she instructed her tenders to tell her husband, "All must die. 'Tis enough that the child lives."

Upon Pocahontas's death, burial rites were hastily arranged. The expedition leaders were not about to miss the tide, and so set sail out of Gravesend within scant hours after she died. The date was March 21, the time of spring equinox, the time when cosmically speaking all things are equal—that time when both Morning Star and Evening Star, the representatives of the grandsons of Nikomis, Ahone, and Ope, are absent. Indeed, considering that the star known as the planet Venus today disappears from view for several days at this time, spring equinox might be the honored anniversary of Sky Woman's fall.

Pocahontas's body was lowered to the vault beneath the chancel floor at Gravesend's church, named for St. George, the dragon slayer, as the ship that bore her to her death had been. Her remains were interred inside the church, as was the practice for people of high status at the time. Entered into the register of burial of the parish of Gravesend we can still read:

−1616−
March 21.—Rebecca Wrolfe wyffe of Thomas Wrolfe gent.
A Virginia Lady borne was buried in the Chauncel.

In 1727 the ancient church, which dated from Saxon times, as its listing in the Domesday Book proves, was destroyed by fire. When it was rebuilt the remains of the princess from a land far, far away were no longer in evidence. In 1897, during an expansion project, some charred skulls and bones were discovered by the construction crew. These were removed and reinterred communally, in a churchyard plot owned by a local family. Thus the exact location of her unmixed remains is unknown, and properly so. A woman who was raised to think that self and community were one and the same could never have rested easy alone.

From the place where thought resides in the head of First Woman, the mind-altering plant of the Americas, *apook,* tobacco, flourished all over the world. In the remains of Pocahontas the spirit of the DNA-altering herb took root in English soil, there at the mouth of the river the Romans named Isis, after the Egyptian mother of gods. The manito and the tradition did not fade, but spread. Altering human consciousness in the old world, from Atlantic to Pacific, it whipped the winds of change, planting the dream, *powa,* in the minds of humans everywhere.

EPILOGUE I: JAMESTOWN COLONY

Rolfe's prospects had risen considerably over the months he spent in England. The high command aboard ship were mostly old friends of his; Hamor, Rolfe's buddy and confidant, was newly named vice-admiral. The new deputy governor and admiral of Virginia, Samuel Argall, was the man who had brought John Smith the news of his dashed ambitions several fateful years earlier, and had, after Smith was out of the picture, triumphantly delivered his royal hostage, Pocahontas, to his allies in Jamestown. In November 1616, at the same board meeting where Hamor and Argall had received their promotions, Rolfe had been promoted too. For his strenuous efforts on behalf of the Virginia Company's bottom line, Rolfe was appointed secretary and recorder-general to the colony of Virginia.

This honor came partly as recognition for his successful work in developing a very marketable product, making a great personal sacrifice in marrying the great Powhatan king's "salvage daughter" and thus bringing peace to the region, and pulling off one of the more resounding promotional tours in English marketing history. It was plain to the executives of the Virginia Company that Rolfe would press on with his notably successful efforts, helping to build a place where men could worship their god free from government interference. Rolfe, who had several land holdings awaiting him in Virginia, was a major player in a growing export enterprise. The investors were hopeful that with the leadership of men such as Rolfe, Argall, and Hamor their enterprise, which had been disappointing until the success of Pocahontas's visit brought renewed interest, would begin to thrive.

EPILOGUE II: THE POWHATANS, PEOPLE OF THE DREAM

Soon after receiving word from Rolfe of Rebecca's death, Wahunsenacawh removed himself—or was removed by the Great Council of Women—from his office as Powhatan, and was replaced by his brother Opechancanough. The fabled Peace of Pocahontas limped on for four more years, but tensions continued to mount between the English and the Powhatans. English violations of agreements, coupled with Opechancanough's strong anti-English sentiments, and intensified by the death of Pocahontas and the loss of most of her Indian retinue, led Opechancanough to marshal what forces he could and attack. Advised and no doubt cheered on by Uttamatamakin, the Powhatans increased their engagements with the enemy. The English were in some disarray, and because they would not farm for food when they could plant a money-making crop, they found themselves once again fallen upon hard times. They were dependent on the Indians for food and other necessities, and their rancor against the Indians grew. At the same time hardships increased for the people of the *tsenacommacah*. About four years after Rolfe and the others had returned to Virginia there was a major battle between the two groups. It resulted in few deaths among the English, but Opechancanough, by now a very weak, old man, withdrew his forces, and the Indian people began to move inland, away from the growing English settlements.

A dozen or so years after the death of the woman who built a *powa* bridge between two worlds, *manito aki* and Faerie, the Native population of the tidewater region was dramatically reduced. Diseases that came with the illegal aliens, bacteriological and viral, psychosocial and spiritual, did their deadly work. As a similar process had decimated the tribal peoples of Europe in the preceding centuries, it decimated those of the eastern seaboard. Soon, almost without notice, the great Powhatan Empire lived only in a small inscription on the engraved portrait of its most famed Beloved Woman.

The Peace of Pocahontas lasted for about five years after her death. It ended the day of Opechancanough's attack on Jamestown. That same day Rolfe followed her to his own grave, and very soon thereafter Wahunsenacawh breathed his last.

CODA: THE BRIDGE

When she left her last words for her husband, presumably Pocahontas was referring to young Thomas, who was ill also. So ill, in fact, that when they reached Plymouth Rolfe felt forced to leave him in the hands of Sir Lewis Stukely, who placed him with his own physician, Dr. Manouri. Rolfe was convinced by his friends Argall and Hamor that it was too dangerous to subject the sick child to the difficulties of crossing the ocean. The boy was eventually transferred to the custody of Rolfe's brother, who lived in London, where he was raised. Thomas never again saw his father, or anyone he had known in the first years of his life. With the exception of Matachanna, the unlucky Powhatan women in Lady Rebecca's retinue were dropped off at the English colony in Bermuda when it anchored there before continuing on to Virginia. Married off to local Indian men on the island, they never returned home. Thomas was raised in England, and though when he entered adolescence he took up residence at his family holdings in Virginia, he refrained from mixing with his Indian kin. It would seem that he lacked his mother's vision and her great gifts of bridging the great gulf between two worlds. What he did contribute was to serve as the first link in a heredity chain of descendants of Pocahontas and John Rolfe. The line was perilously close to dying out until the lifetime of their great-grandson, Col. John Bolling. The Rolfe descendants were among the earliest of Virginia's first families. The mixed-blood descendants of Powhatan-English stock number around 3 million at the present time. They reside in the United States, England, and, it is likely, all over the world. It is to be hoped that some of them dream.

Appendix 1

John Rolfe's Letter to Master Thomas Dale

The coppie of the Gentle-mans letters to Sir Thomas Dale, that after married Powhatans daughter, containing the reasons moving him thereunto.

Honourable Sir, and most worthy Governor:

When your leasure shall best serve you to peruse these lines, I trust in God, the beginning will not strike you into a greater admiration, then the end will give you good content. It is a matter of no small moment, concerning my own particular, which here I impart unto you, and which toucheth mee so neerely, as the tendernesse of my salvation. Howbeit I freely subject my selfe to your grave and mature judgement, deliberation, approbation, and determination; assuring my selfe of your zealous admonitions, and godly comforts, either perswading me to desist, or incouraging me to persist therin, with a religious and godly care, for which (from the very instant, that this began to roote it selfe within the secret bosome of my brest) my daily and earnest praiers have bin, still are, and ever shall be produced forth with as sincere a godly zeale as I possibly may to be directed, aided and governed in all my thoughts, words, and deedes, to the glory of God, and for my eternal consolation. To persevere wherein I never had more neede, nor (till now) could ever imagine to have bin moved with the like occasion.

But (my case standing as it doth) what better worldly refuge can I here seeke, then to shelter my selfe under the safety of your favourable protection? And did not my ease proceede from an unspotted conscience, I should not dare to offer to your view and approved judgement, these passions of my troubled soule, so full of feare and trembling in hypocrisie and dissimulation. But knowing my owne innocency and godly fervor, in the whole prosecution hereof, I doubt not of your benigne acceptance, and clement construction. As for malicious depravers, and turbulent spirits, to whom nothing is tastful but what pleaseth their unsavory paalat, I passe not for them being well assured in my perswasion (by the often trial and proving of my selfe, in my holiest meditations and praiers) that I am called hereunto by the spirit of God; and it shall be sufficient for me to be protected by your selfe in all vertuous and pious indevours. And for my more happie proceeding herein, my daily oblations shall ever be addressed to bring to passe so good effects, that your selfe, and all the world may truely say: This is the worke of God, and it is marvelous in our eies.

But to avoid tedious preambles, and to come neerer the matter: first suffer me with your patence, to sweepe and make cleane the way wherein I walke, from all suspicions and doubts, which may be covered therein, and faithfully to reveale unto you, what should move me hereunto.

Let therefore this my well advised protestation, which here I make betweene God and my own conscience, be a sufficient witnesse, at the dreadfull day of judgement (when the secret of all mens harts shall be opened) to condemne me herein, if my chiefest intent and purpose be not, to strive with all my power of body and minde, in the undertaking of so mightie a matter, no way led (so farre forth as mans weakenesse may permit) with the unbridled desire of carnall affection: but for the good of this plantation, for the honour of our countrie, for the glory of God, for my owne salvation, and for the converting to the true knowledge of God and Jesus Christ, an unbeleeving creature, namely Pokahuntas. To whom my hartie and best thoughts are, and have a long time bin so intagled, and inthralled in so intricate a laborinth, that I was even awearied to unwinde my selfe thereout. But almighty God, who never faileth his, that truly invocate his holy name hath opened the gate, and led me by the hand that I might plainely see and discerne the safe paths wherein to treade.

To you therefore (most noble Sir) the patron and Father of us in this countrey doe I utter the effects of this setled and long continued affection (which hath made a mightie warre in my mediations) and here I doe truely relate, to what issue this dangerous combate is come unto, wherein I have not onely examined, but throughly tried and pared my thoughts even to the quick, before I could send and fit wholesome and apt applications to cure so daungerous an ulcer. I never failed to offer my daily and faithfull praiers to God, for his sacred and holy assistance. I forgot not to set before mine eies the frailty of mankinde, his prones to evill, his indulgencie of wicked thoughts, with many other imperfections wherein man is daily insnared, and oftentimes overthrowne, and them compared to my present estate. Nor was I ignorant of the heavie displeasure which almightie God conceived against the sonnes of Levie and Israel for marrying strange wives, nor of the inconveniences which may thereby arise, with other the like good motions which made me looke about warily and with good circumspection, into the grounds and principall agitations, which thus should provoke me to be in love with one whose education hath bin rude, her manners barbarous, her generation accursed, and so discrepant in all nurtriture frome my selfe, that oftentimes with feare and trembling, I have ended my private controversie with this: surely these are wicked instigations, hatched by him who seeketh and delighteth in mans destruction; and so with fervent praiers to be ever preserved from such diabolical assaults (as I tooke those to be) I have taken some rest.

Thus—when I had thought I had obtained my peace and quitnesse, beholde another, but more gracious tentation hath made breaches into my holiest and strongest meditations; with which I have bin put to a new traill, in a straighter manner then the former: for besides the many passions and sufferings which I have daily, hourely, yea and in my sleepe indured, even awaking mee to astonishment, taxing mee with remisnesse, and carlesnesse, refusing and neglecting to performe the duetie of a good Christian, pulling me by the eare, and crying:

why dost not thou indevour to make her a Christian? And these have happened to my greater wonder, ven when she hath bin furthest seperated from me, which in common reason (were it not an undoubted worke of God) might breede forgetfulnesse of a farre more worthie creature. Besides, I say the holy spirit of God often demaunded of me, why I was created?

If not for transitory pleasures and worldly vanities, but to labour in the Lords vineyard, there to sow and plant, to nourish and increase the fruites thereof, daily adding witt the good husband in the Gospell, somewhat to the tallent, that in the end the fruites may be reaped, to the comfort of the laborer in this life, and his salvation in the world to come? And if this be, as undoubtedly this is, the service Jesus Christ requireth of his best servant: wo unto him that hath these instruments of pietie put into his hands and wilfillly despiseth to worke with them. Likewise, adding hereunto her great apparance of love to me, her desire to be taught and instructed in the knowledge of God, her capablenesse of understanding, her aptnesse and willingnesse to receive anie good impression, and also the spirituall, besides her owne incitements stirring me up hereunto.

What should I doe? Shall I be of so untoward a disposition, as to refuse to leade the blind into the right way? Shall I be so unnaturall, as not to give bread to the hungrie? or uncharitable, as not to cover the naked? Shall I despise to actuatethese pious dueties of a Christian? Shall the base feare of displeasing the world, overpower and with holde mee from revealing unto man these spirituall workes of the Lord, which in my meditations and praiers, I have daily made knowne unto him? God forbid. I assuredly trust hee hath thus delt with me for my eternall felicitie, and for his glorie: and I hope so to be guided by his heavenly graice, that in the end by my faithfilll paines, and christianlike labour, I shall attaine to that blessed promise, Pronounced by that holy Prophet Daniell unto the righteous that bring many unto the knowledge of God. Namely, that they shall shine like the starres forever and ever. A sweeter comfort cannot be to a true Christian, nor a greater incouragement for him to labour all the daies of his life, in the performance thereof, nor a greater gaine of consolation, to be desired at the hower of death, and in the day of judgement.

Againe by my reading, and conference with honest and religious persons, have I received no small encouragement, besides serena mea conscientia, the cleerenesse of my conscience, clean from the filth of impurity, quoe est instar muri ahenei, which is unto me, as a brasen wall. If I should set down at large, the perhioations and godly motions, which have striven within mee, I should but make a tedious and unnecessary volume. But I doubt not these shall be sufficient both to certifie you of my tru intents, in discharging of my dutie to God, and to your selfe, to whose gracious providence I humbly submit my selfe, for his glory, your honour, our Countreys good, the benefit of this Plantation, and for the converting of one unregenerate, to regeneration; which I beseech God to graunt, for his deere Sonne Christ Jesus his sake.

Now if the vulgar sort, who square all mens actions by the base rule of their owne filthinesse, shall taxe or taunt me in this my godly labour: let them know, it is not any hungry appetite, to gorge my selfe with incontinency; sure (if I would, and were so sensually inclined) I might satisfie such desire, though not without a seared conscience, yet with Christians more pleasing to the eie, and lesse fearefull in the offence unlawfully committed. Nor am I in so desperate an estate, that I regard not what becommeth of mee; nor am I out of hope but one day to see my Country, nor so void of friends, nor mean in birth, but there to obtain a mach to my great content: nor have I ignorantly passed over my hopes there, or regardlesly seek to loose the love of my friends, by taking this course: I know them all, and have not rashly overslipped any.

But shal it please God thus to dispose of me (which I earnestly desire to fulfill my ends before sette down) I will heartely accept of it as a godly taxe appointed me, and I will never cease, (God assisting me) untill I have accomplished, and brought to perfection so holy a worke, in which I will daily pray God to blesse me, to mine, and her eternall happines. And thus desiring no longer to live, to enjoy the blessings of God, then this my resolution doth tend to such godly ends, as are by me before declared: not doubting of your favourable acceptance, I take my leave, beseeching Almighty God to raine downe upon you, such plenitude of his heavenly graces, as your heart can wish and desire, and so I rest,

At your command most willing to be disposed of

John Rolfe[1]

John Smith's Letter to Queen Anne

To the most high and virtuous princess, Queen Anne of Great Britain
   Most admired Queen,
   The love I bear my God, my King and country, hath so oft em-
boldened me in the worst of extreme dangers, that now honesty doth
constrain me to presume thus far beyond myself, to present your
Majesty this short discourse: if ingratitude be a deadly poison to all
honest virtues, I must be guilty of that crime if I should omit any
means to be thankful.

   So it is, that some ten years ago being in Virginia, and taken prisoner by the power of Powhatan their chief King, I received from
this great Salvage exceeding great courtesy, especially from his son
Nantaquaus, the most manliest, comeliest, boldest spirit, I ever saw in
a Salvage, and his sister Pocahontas, the Kings most dear and well-
beloved daughter, being but a child of twelve or thirteen years of age,
whose compassionate pitiful heart, of my desperate estate, gave me
much cause to respect her: I being the first Christian this proud King
and his grim attendants ever saw: and thus enthralled in their barbarous power, I cannot say I felt the least occasion of want that was in
the power of those my mortal foes to prevent, notwithstanding all
their threats. After some six weeks fatting amongst those Salvage
courtiers, at the minute of my execution, she hazarded the beating
out of her own brains to save mine; and not only that, but so prevailed
with her father, that I was safely conducted to Jamestown: where I

found about eight and thirty miserable poor and sick creatures, to keep possession of all those large territories of Virginia; such was the weakness of this poor commonwealth, as had the salvages not fed us, we directly had starved. And this relief, most gracious Queen, was commonly brought us by this Lady Pocahontas.

Notwithstanding all these passages, when inconstant fortune turned our peace to war, this tender virgin would still not spare to dare to visit us, and by her our jars have been oft appeased, and our wants still supplied; were it the policy of her father thus to employ her, or the ordinance of God thus to make her his instrument, or her extraordinary affection to our nation, I know not: but of this I am sure; when her father with the utmost of his policy and power, sought to surprise me, having but eighteen with me, the dark night could not affright her from coming through the irksome woods, and with watered eyes gave me intelligence, with her best advice to escape his fury; which had he known, he had surely slain her.

Jamestown with her wild train she as freely frequented, as her fathers habitation; and during the time of two or three years, she next under God, was still the instrument to preserve this colony from death, famine and utter confusion; which if in those times, had once been dissolved, Virginia might have lain as it was at our first arrival to this day.

Since then, this business having been turned and varied by many accidents from that I left it at: it is most certain, after a long and troublesome war after my departure, betwixt her father and our colony; all which time she was not heard of.

About two years after she herself was taken prisoner, being so detained near two years longer, the colony by that means was relieved, peace concluded; and at last rejecting her barbarous condition, she was married to an English Gentleman, with whom at this present she is in England; the first Christian ever of that Nation, the first Virginian ever spoke English, or had a child in marriage by an Englishman: a matter surely, if my meaning be truly considered and well understood, worthy a Princes understanding.

Thus, most gracious Lady, I have related to your Majesty, what at your best leisure our approved Histories will account you at large, and done in the time of your Majesty's life; and however this might be presented you from a more worthy pen, it cannot from a more honest heart, as yet I never begged anything of the state, or any: and it is my want of ability and her exceeding desert; your birth, means, and authority; her birth, virtue, want and simplicity, doth make me thus bold, humbly to beseech your Majesty to take this knowledge of her, though it be from one so unworthy to be the reporter, as myself, her husbands estate not being able to make her fit to attend your Majesty. The most and least I can do, is to tell you this, because none so oft hath tried it as myself, and the rather being of so great a spirit, however her stature: if she should not be well received, seeing this Kingdom may rightly have a Kingdom by her means; her present love to us and Christianity might turn to such scorn and fury, as to divert all this good to the worst of evil; whereas finding so great a Queen should do her some honor more than she can imagine, for being so kind to your servants and subjects, would so ravish her with content, as endear her dearest blood to effect that, your Majesty and all the Kings honest subjects most earnestly desire.

And so I humbly kiss your gracious hands,

Captain John Smith, 1616

Don Diego de Molina's Letter to King Philip of Spain, 1613

The person who will give this to Your Lordship is very trustworthy and Your Lordship can give credence to everything he will say, so I will not be prolix in this but will tell in it what is most important. Although with my capture and the extraordinary occurrences following it His Majesty will have opened his eyes and seen this new Algiers of America, which is coming into existence here, I do not wonder that in all this time he has not remedied it because to effect the release would require an expedition, particularly as he lacks full information for making a decision. However I believe that with the aid of Your Lordship's intelligence and with the coming of the caravel[1] to Spain, His Majesty will have been able to determine what is most important and that that is to stop the progress of a hydra in its infancy, because it is clear that its intention is to grow and encompass the destruction of all the West, as well by sea as by land and that great results will follow I do not doubt, because the advantages of this place make it very suitable for a gathering-place of all the pirates of Europe, where they will be well received. For this nation has great thoughts of an alliance with them. And this nation by itself will be very powerful because as soon as an abundance of wheat shall have been planted and there shall be enough cattle, there will not be a man of any sort whatever who will not alone or in company with others fit out a ship to come here and join the rest, because as Your Lordship knows this Kingdom abounds in poor people who abhor peace—and of necessity because in peace they perish—and the rich are so greedy and selfish that they even cherish a desire for the Indies and the gold and silver there—notwithstanding that there will not be much lack of these here, for they have discovered some mines which are considered good, although they have not yet been able to derive profit from them. But when once the preliminary steps are taken there are many indications that they will find a large number in the mountains. So the Indians say and they offer to show the locations that they know and they say that near the sources of the rivers, as they come down from the mountain, there is a great quantity of grains of silver and gold, but, as they do not set any value on these but only on copper which they esteem highly, they do not gather them.

As yet these men have not been able to go to discover these although they greatly desire it, nor to pass over this range to New Mexico and from there to the South Sea where they expect to establish great colonies and fit out fleets with which to become lords of that sea as well as of this, by colonizing certain islands among those to the east of the channel of Bahama and even to conquer others, as Puerto Rico, Santo Domingo and Cuba.[2] And although this would be difficult at the least, we have already seen signs of these purposes in the colonizing of Bermuda where they are said to have strong fortifications, because the lay of the land is such that a few can defend themselves against a great number and prevent disembarking and landing.[3] The depth as I have understood is not enough for ships of a hundred tons, but I believe that they make it out less than it is, for that island has already been described in the relation of Captain Diego Ramirez who was stranded there, and it seems to me that larger vessels can enter. I do not recall it well but the description is in the house of Don Rodrigo de Aguiar of the Council of the Indies and the register is in Seville in the house of the licentiate Antonio Moreno, cosmographer of the Council. But above all this captain will give enough information of the island, and it is very important for the military actions which may have to take place in it. Its fertility is great, there is abundance of fish and game, and pork as much as they can want, and so they get along very well in that colony because they have little need of

England, for they are likewise rich in amber and pearl of which in a very few months it is said they have sent to that kingdom more than fifty thousand ducats in value, reckoning the ounce at a moderate price. Four days ago a vessel arrived here that brought them men and provisions and they do not cease talking of the excellence of that island and its advantages.

The soil in this place is very fertile for all species, only not for those which require much heat, because it is cold. There is much game and fish, but as they have not begun to get profit from the mines, but only from timber, the merchants have not been able to maintain this colony with as much liberality as was needed and so the people have suffered much want, living on miserable rations of oats or maize and dressing poorly. For which reason, if today three hundred men should come, this same year would destroy more than one hundred and fifty, and there is not a year when half do not die. Last year there were seven hundred people and not three hundred and fifty remain, because little food and much labor on public works kills them and, more than all, the discontent in which they live seeing themselves treated as slaves with cruelty.[4] Wherefore many have gone over to the Indians, at whose hands some have been killed, while others have gone out to sea, being sent to fish, and those who remain have become violent and are desirous that a fleet should come from Spain to take them out of their misery. Wherefore they cry to God of the injury that they receive and they appeal to His Majesty in whom they have great confidence, and should a fleet come to give them passage to that kingdom, not a single person would take up arms.[5] Sooner would they forfeit their respect and obedience to their rulers who think to maintain this place till death.

And although it is understood there that the merchants[6] are deserting this colony, this is false for it is a stratagem with which they think to render His Majesty careless, giving him to understand that this affair will settle itself, and that thus he will not need to go to the expense of any fleet whatever to come here. With eight hundred or one thousand soldiers he could reduce this place with great ease, or even with five hundred, because there is no expectation of aid from

England for resistance and the forts which they have are of boards and so weak that a kick would break them down, and once arrived at the ramparts those without would have the advantage over those within because its beams and loopholes are common to both parts—a fortification without skill and made by unskilled men. Nor are they efficient soldiers, although the rulers and captains make a great profession of this because of the time they have served in Flanders on the side of Holland, where some have companies and castles. The men are poorly drilled and not prepared for military action.

However they have placed their hope on one of two settlements, one which they have founded twenty leagues up the river in a bend on a rugged peninsula[7] with a narrow entrance by land and they are persuaded that there they can defend themselves against the whole world. I have not seen it but I know that it is similar to the others and that one night the Indians entered it and ran all over the place without meeting any resistance, shooting their arrows through all the doors, so that I do not feel that there would be any difficulty in taking it or the one in Bermuda, particularly if my advice be taken in both matters as that of a man who has been here two years and has considered the case with care. I am awaiting His Majesty's decision and am desirous of being of some service and I do not make much of my imprisonment nor of the hardships which I have suffered in it, with hunger, want and illness, because one who does a labor of love holds lightly all his afflictions. The ensign Marco Antonio Perez died fifteen months ago, more from hunger than illness, but assuredly with the patience of a saint and the spirit of a good soldier. I have not fared very ill, but tolerably so, because since I arrived I have been in favor with these people and they have shown me friendship as far as their own wretchedness would allow, but with genuine good-will. The sailor who came with me is said to be English and a pilot. He declares that he is from Aragon and in truth no one would take him for a foreigner.[8]

This country is located in thirty-seven and a third degrees, in which is also the bay which they call Santa Maria.[9] Five rivers empty into this, very wide and of great depth—this one at its entrance nine fathoms and five and six within. The others measure seven, eight and twelve; the bay is eight leagues at its mouth but in places it is very wide, even thirty leagues.[10] There is much oak timber and facilities for making ships, trees for them according to their wish—very dark walnut which they esteem highly and many other kinds of trees.

The bearer is a very honorable Venetian gentleman, who having fallen into some great and serious errors is now returned to his first religion and he says that God has made me his instrument in this, for which I give thanks. He wishes to go to Spain to do penance for his sins. If I get my liberty I think of helping him in everything as far as I shall be able. I beseech Your Lordship to do me the favor of making him some present, for I hold it certain that it will be a kindness very acceptable to our Lord to see in Your Lordship indications that charity has not died out in Spain. And so Your Lordship ought to have charity and practise it in the case of a man who goes from here poor and sick and cannot make use of his abilities, and if I have to stay here long I am no less in need of Your Lordship's help (as you will learn from the report of this man, who will tell you how I am faring). Your Lordship might aid me by sending some shipstores such as are brought here for certain private individuals and in particular cloth and linen for clothing ourselves (this man and me) because we go naked or so ragged that it amounts to the same, without changing our shirts in a month, because, as the soldier says, my shirts are an odd number and do not come to three. I trust in God who will surely help me since He is beginning to give me my health which for eleven months has failed me. I have not sufficient opportunity to write to His Majesty. Your Lordship will be able to do this giving him notice of everything I am telling. May God guard Your Lordship as I desire. From Virginia, May 28 (according to Spanish reckoning), 1613.

If Your Lordship had the key to my cipher, I should write in it. But this letter is sewed between the soles of a shoe, so that I trust in God that I shall not have done wrong in writing in this way. When I first came here I wrote His Majesty a letter which had need of some interpretation and directed it with others to Your Lordship. I do not know whether you have received them.

I thought to be able to make a description of this country but the publicity of my position does not give me opportunity for it, but that which is most to the point is that the bay runs northeast by east and at four leagues distance from its mouth is this river from the south, nine fathoms in depth. At the entrance is a fort[11] or, to speak more exactly, a weak structure of boards ten hands high with twenty-five soldiers and four iron pieces. Half a league off is another[12] smaller with fifteen soldiers, without artillery. There is another[13] smaller than either half a league inland from here for a defence against the Indians. This has fifteen more soldiers. Twenty leagues off is this colony[14] with one hundred and fifty persons and six pieces; another twenty leagues further up is another colony[15] strongly located—to which they will all betake themselves if occasion arises, because on this they place their hopes—where are one hundred more persons and among them as here there are women, children and field laborers, which leaves not quite two hundred active men and those poorly disciplined.[16]

Notes

OO-MAA'O / INTRODUCTION

1. At the beginning of the seventeenth century the kind of materialist world we live in was in its infancy; nevertheless, the idea that the real was that which could be perceived by the physical senses and/or was that which was authorized by Church and/or State had gained a firm footing. When I refer to "modern people"—meaning most of us living in the postindustrial world, either in emergent or older nations—I am referring to people who hold this constellation of beliefs and assumptions about the nature of reality. While for many today the authorization comes from academics, bureaucracies, or other sources of "recognized fact," the principle remains pretty much the same.

2. That means that married couples lived in the woman's mother's village or extended-family dwelling, lineage was traced through the mother's line, and fields, houses, and most material items belonged to the women, as did the harvest of field or hunt.

3. He spelled his name Ralegh rather than Raleigh as it has commonly been misspelled. Thomas Hariot was much involved in the exploration and development of English interests in America. It was he who instructed John Smith in the Algonquin he himself had learned on an earlier trip—to Roanoke. He was an Enochian.

4. Enochian magic shares some of its teachings with the cabalistic system. It was originally presented by Sir John Dee, the English court astrologer to Queen Elizabeth in the sixteenth century. Dee, working with Edward Kelley, learned of the Enochian system when it was "revealed" to him by the Enochian angels that inhabited the Watchtowers and the Aethyrs—the subtle regions of the universe. While Kelley looked deeply into a crystal "shewstone," describing aloud what he saw, Dee took notes. He then developed a series of runes to correspond to the Enochian alphabet and language. The Enochian alphabet has twenty-one magical letters and is the language "of the angels." The runes were placed on a series of square tablets that corresponded to the planes of the magical universe. The four planes, or Watchtowers, correspond to the elements of earth, air, fire, and water. Each of these represents aspects of our world or relevance to living things; their correspondences form the basis of Enochian workings. The other important aspect of Enochian magic is that it describes the thirty Aethyrs, zones that penetrate the Watchtowers. The Enochian magician learns to leave his or her physical body and travel through these Aethyrs in his/her body of light, in order to become a living embodiment of knowledge and power. The Enochian system is very simple to use, and is very powerful as well. http://www.mtsn.org.uk/acdepts/english/tempest/magic.htm. May 1, 2003.

5. Not an easy process, it continues; while much of the human world has heard of "American democracy," many regions are about as far from making it a reality for their country as those first Englishmen in America were—maybe further.

CHAPTER 1. *APOWA* / DREAM-VISION

1. William Strachey, *Historie of Travell into Virginia Britania* (1612; reprint, ed. Louis B. Wright and Virginia Freund, Hakluyt Society, 1953).

2. The matter of clans, bands, tribes, nations, moieties, marriage laws, and the like is a complex and well-researched topic. I can give only a very general idea of these matters as they relate to Pocahontas here.

3. Taken from the Greek, the word *myth* extends back to earliest records of Indo-European. It means "to make," but to do so in a magical or sacred manner rather than in a purely physical one.

4. F. David Peat, *Blackfoot Physics: A Journey into the Native American Universe* (London: Fourth Estate, 1994), 87–88.

5. Frances Mossiker, *Pocahontas: The Life and the Legend* (New York: De Capo Press, 1976), 22.

6. Peat, *Blackfoot Physics,* 80.

7. Similarly, words containing the phoneme *ma,* such as *mamanantowick* and *manito,* signified a greater degree of *powa,* or a deeper connection to what physicist David Bohm called "the implicate order."

8. One might speculate that the tribe's name was composed of power-associated morphemes, "po," "ma," "ni," and "ta" forming "mani t(a)-po(w)," or "manito-powa," suggesting a great deal of inherent *powa* indeed. The way of making meaning in Indian languages has a lot to do with rearranging meaningful syllables, or morphemes, so that several implications, including person, case, tense, and number, are accounted for. For more on this see Roger Williams, *A Key into the Language of America: or, An help to the Language of the Natives in that part of America, called New-England* (Bedford, MA: Applewood Books, year unknown [1643; 1936]); and Frederic W. Gleach, *Powhatan's World and Colonial Virginia: A Conflict of Cultures* (Lincoln: Univ. of Nebraska Press, 1997).

9. Ronald Goodman, *Lakota Star Knowledge: Studies in Lakota Stellar Theology* (Rosebud Sioux Reservation: Sinte Gleska Univ. Press, 1992), 1.

10. "The stenographic record notes that the [sacred] pipe was brought by the White Buffalo Woman, '800 years ago way back east.' " Raymond DeMallie, ed., *The Sixth Grandfather: Black Elk's Teachings Given to John G. Neihardt* (Lincoln: Bison Books, Univ. of Nebraska Press, 1984), 282 n.

11. Betty Donahue, telephone interview by author, November 2002.

CHAPTER 2. *POCAHONTAS* / MISCHIEF

1. Strachey, *Historie.* A writer by inclination, perhaps one of those engaged in the new enterprise of journalism, he associated with the London glitterati, such as Christopher Marlowe, Ben Jonson, and William Shakespeare, all court-patronized writers. Heady company indeed.

2. John Smith, *The Generall Historie of Virginia, New England, and the Summer Isles.*

3. ———.

4. Grace Steele Woodward, *Pocahontas* (Norman: Univ. of Oklahoma Press, 1969), 92.

5. Mossiker, *Pocahontas,* 139, quoting Symonds, source uncited.

6. Charles R. Larson, *American Indian Fiction* (Albuquerque: Univ. of New Mexico Press, 1978), 24.

7. William Brandon, *The Last Americans: The Indian in American Culture* (New York: McGraw-Hill, 1974), 100.

8. ————., 101.

9. http://www.jbtank.com/indians/saca.html (5 April 2003).

10. John Fire Lame Deer and Richard Erdoes, *Lame Deer: Seeker of Visions* (New York: Simon and Schuster, 1972), 114–15.

11. Peat, *Blackfoot Physics,* 140.

12. A Cochiti Pueblo story.

13. An anonymous story, maybe Oklahoma Lenape, based on John Bierhorst, ed., *The White Deer and Other Stories Told by the Lenape* (New York: William Morrow, 1995), 35–37.

14. *Reiterated*—rather than *replicated*—may be the more accurate term here.

15. Indigenous traditional narratives include the "X-factor" in the figure of the trickster, whether it be coyote, rabbit, spider-boy, fox, mouse, raven, goddess, god, or court-jester, the holy fool.

16. Charles deLint, *The Little Country* (New York: William Morrow, 1991), 280.

17. Smith describes his ambitions to remain in Virginia in a section appended to book two of his *Generall Historie.*

18. Woodward, *Pocahontas,* 113.

19. Larson, *American Indian Fiction,* 25.

20. The "Stockholm syndrome" is a name given to a phenomenon in which one who is taken hostage identifies with the captors to the point that she or he becomes their ally, as was suggested in the case of Hearst Publishing magnate Randolph Hearst's daughter Patricia Hearst, abducted by a group that called itself the Symbionese Liberation Army. Hearst was abducted in the early 1970s. The Stockholm syndrome is so named because of just such a hostage incident that took place in Stockholm, Sweden.

CHAPTER 3. *MANITO AKI* / FAERIE

1. The Oral Tradition, while a major area of study for ethnographers and other students of American Indian life, is not confined to the Americas. As a bit of terminology, it refers to traditional narrative and ceremonial material kept by a people, whether tribal, peasant, or urban. Technically, of course, there is no Native American Oral Tradition because there is no such place or nation as Native America. Each Native Nation—Powhatan, Navajo, Lakota, Cherokee, Hawaiian, Samoan, Micronesian, Filipino, Egyptian, Meso-potamian, Greek, English, and so forth—has one or more Oral Traditions.

2. There is an eerie similarity in the mythologies of the world that goes unaccounted for by scholars; the closest explanation may be Carl Jung's idea of archetypes, in which the Swiss psychoanalyst theorized that there are certain characters common to human consciousness that are somehow embedded in the human brain. Alternatively, he thought, perhaps these figures were embedded in what he, an alchemist as well as a psychoanalyst, referred to as the *Spiritus Mundi,* the Spirit-Body of Earth. *Spiritus Mundi* can be thought of as a European version of *manito aki.* Both are seen as a space beyond the tangible or physical in which a variety of beings or at least forces exist. The argument seems to be more about whether the *Spiritus Mundi* is a collective unconscious generated by humans over millennia or if it exists independent of human thought. On the other

hand, it may be that these correspondences exist because the implicate order works in particular ways, hosts particular entities with particular characteristics, and in general operates like the other worlds humans have taught their young about and studied for entire lifetimes.

3. *Sir Gawain and the Green Knight,* trans. Jessie L. Weston (Gawain Menu, University of Rochester, the Camelot Project; http://www.lib.rochester.edu/camelot/sggk.htm, 24 September 2002).

4. Williams, *A Key into the Language of America,* p. 126.

5. ———., 126.

6. Melissa Jane Fawcett (Tantaquidgeon) and Joseph Bruchac, *Makiawisug: The Gift of the Little People* (Mohegan Nation: Little People Publications, 1997). In their dedication Fawcett (Mohegan) and Bruchac (Abenaki) thank Flying Bird / Jeets Bodernasha, "for passing on the story." They inform us that Mrs. Bodernasha got it from her grandmother, Nonner Martha, and later passed it on to Gladys Tuntaquidgeon, Melissa's grandmother. Mrs. Tuntaquidgeon is Medicine Woman of the Mohegan people today, and the provenance of the narrative establishes it securely in the Oral Tradition of their Algonquin people.

7. ———., 1.

8. ———., 18.

9. Mossiker, *Pocahontas,* 161, 363 n.

10. Rafe (Ralph) Hamor, "A True Discourse of the Present Estate of Virginia," in *Jamestown Narratives: Eyewitness Accounts of the Virginia Colony: The First Decade: 1607–1617,* ed. Edward Wright Haile (Champlain, VA: Round House Press, 1998), 827. This book includes delightful commentary by Haile, who uses versions rendered in 1860 and 1957 for his version. In context, at that time *reduced* could mean "restored back" or even "constructed."

11. Charles deLint, *Moonlight and Visions* (New York: Tor Books, Tom Doherty Associates, 1999), 325.

12. Joseph Campbell, *The Inner Reaches of Outer Space* (New York: Harper & Row, 1986), 17.

13. Hamor, "A True Discourse," 803.

14. Mossiker, *Pocahontas,* 159–63.

15. Five years before Charles I, the son and heir of James II, was beheaded in 1649 by the Puritan faction, the theaters were destroyed and stage plays banned.

16. *The Tempest,* 5.1.182–84.

17. William M. Kelso et al., *Jamestown Rediscovery: The Search for the 1607 James Fort* (Richmond: Association for the Preservation of Virginia Antiquities, 1995–2001), 3:8.

18. ———., 3:8.

19. ———., 4:23.

20. DeLint, *Moonlight and Visions,* 453.

21. Haile, *Jamestown Narratives,* 793.

22. The version I am basing my account on is recorded in Victor Barnouw, ed., *Wisconsin Chippewa Myths and Tales and Their Relation to Chippewa Life* (Madison: Univ. of Wisconsin Press, 1986). Chippewa and Ojibwaj are closely related, and sometimes the names are used interchangeably.

23. Peat, *Blackfoot Physics,* 140. One method of mixing omelets re-creates the effect as well. The process, whether described in abstract or practical terms, closely resembles the Julia Set devised by mathematician Benoit Mandelbrot, along with his astonishing Mandelbrot Set, which itself resembles nothing so much as the Venus of Düsseldorf to my eye.

24. Helen C. Rountree, *The Powhatan Indians of Virginia, Their Traditional Culture* (Norman: Univ. of Oklahoma Press, 1989), 134. Rountree cites George Percy, *A Trewe Relation* (1612), reprinted in *Tyler's Quarterly* 3 (1921–22): 272.

25. Paula Gunn Allen, *The Sacred Hoop: Recovering the Feminine in American Indian Traditions* (Boston: Beacon Press, 1987); Paula Gunn Allen, *Off the Reservation: Reflections of Boundary-Busting, Border-Crossing Loose Canons* (Boston: Beacon Press, 1999).

26. David Bohm, *Wholeness and the Implicate Order* (London and Boston: Routledge & Kegan Paul, 1981), 178–79.

27. Mossiker, *Pocahontas,* 167.

28. Samuel Lewis, *A Topological Dictionary of England* (1831; http://privatewww.essex. ac.uk/~alan/family/N-StJohnsCam.html; 28 September 2002). To give an idea of the exalted company in which the Reverend Whitaker, the son, had traveled before he took up his mission in the colonies, Lewis lists the following eminent members of the faculty: Roger Ascham; Sir John Cheke; Sir Thomas Wyat; Lord Treasurer Burleigh; Lord Keeper Williams; Dr. John Dee; Thomas Wentworth, Earl of Stafford; Lord Falkland; Dr. William Whitaker; Dr. William Cave; Archbishop Williams; Bishops Day, Gauden, Gunning, Stillingfleet, and Beveridge; Dr. Jenkins, who wrote on the reasonableness of Christianity; Dr. Powell; Dr. Balguy; Dr. Ogden; Thomas Stackhouse, author of the *History of the Bible;* Dr. William Wotton, Dr. Bentley, and Dr. Taylor, the critics; Ben Jonson; the poets John Cleland, Ambrose Philips, Prior, Otway, Broome, Hammond, and Mason; Martin Lister, the naturalist; Francis Peck and Thomas Baker, the antiquaries; and the late Dr. Heberden.

29. Dr. John Dee was a "Man of science and magus extraordinary for two decades England's leading mathematician, it is only in recent years that John Dee's reputation has begun to properly recover from the obloquy attached by an age of militant rationalism to those notorious angel raising episodes in which he engaged in the 1580's. . . . What I am seeking to identify is the political and religious significance of these episodes and the clues they give to the secret society culture of the late Elizabethans." Ron Heisler, "John Dee and the Secret Societies" *Hermetic Journal: 1992;* http://www.levity.com/alchemy/h_dee.html.

30. Haile, *Jamestown Narratives,* 37.

31. Mossiker, *Pocahontas,* 146.

32. ———., 146 n. 10 (citing George Percy).

33. ———., 172.

34. Ann Marie Plane, *Colonial Intimacies: Indian Marriage in Early New England* (Ithaca, NY: Cornell Univ. Press, 2000), 5.

35. Campbell, *Inner Reaches of Outer Space,* 59.

36. Hamor, "A True Discourse," 793.

37. Ivor Noël Hulme, *The Virginia Adventure, Roanoke to James Towne: An Archeological and Historical Odyssey* (Charlottesville: Univ. of Virginia Press, 1994), 328.

38. "The land" John Winthrop refers to is, of course, land that is populated by northeastern Algonquins—the Wampanoags. In this particular case he refers to the land of those whose villages will be used by the English while they build their own. (http://www.nv.cc.va.us/home/nvsageh/Hist121/Part1/winthrop.htm; 4 April 2003).

CHAPTER 4. *APOOK* / THE ESTEEMED WEED

1. Natalie Curtis, recorder and editor, *The Indians' Book: Songs and Legends of the American Indian* (New York: Dover Publications, 1950; reprint, New York: Bonanza Books, Crown Books, 1987), 4–6.

2. ———., 6–8. A version of the myth of First Woman was first published in a collection of Abanaki stories recorded by a Penobscot Algonquin man named Joseph Nicolar in 1893, but it sounds a lot like a lyrical version of earth scientists' account of the development of life on this planet to my ear.

3. Eliphas Levi, *Transcendental Magic, Its Doctrine and Ritual,* trans. Arthur Edward Waite (London: Rider and Company, 1896; reprint, New York: Samuel Weiser, 1970).

4. Iain Gately, *Tobacco: A Cultural History of How an Exotic Plant Seduced Civilization* (New York: Grove Press, 2001), 71.

5. ———., 74.

6. Perhaps, since the publication of Chippewa—that is, western Algonquin (more properly named Anishinabeg)—writer Louise Erdrich's *Love Medicine* and its companion novels, researchers can entertain other, more likely, possibilities. The growing torrent of fiction, poetry, and scholarship by American Indian poets and writers reveals the alternate narrative in ever more telling detail. This, in turn, might lead students of American history to take a look at old lore of various branches of the great Algonquin.

7. Virgil J. Vogel, *American Indian Medicine* (Norman: Univ. of Oklahoma Press, 1990); Jack Weatherford, *Indian Givers: How the Indians of the Americas Transformed the World* (New York: Crown Publishers), 1988.

8. Warren M. Billings, *Jamestown and the Founding of the Nation*, Colonial National Historical Park and Eastern National Park and Monument Association (Gettysburg: Thomas Publications, year unknown), 45.

9. Jeremy Narby, *The Cosmic Serpent: DNA and the Origins of Knowledge* (New York: Putnam, 1998), 156.

10. ———., 42.

11. Mossiker, *Pocahontas,* 172.

12. ———., 149.

13. Gately, *Tobacco,* 6.

14. Rountree, *The Powhatan Indians of Virginia,* 82.

15. ———., 82.

16. ———., 82.

17. Jack Frederick Kilpatrick and Anna Gritts Kilpatrick, *Walk in Your Soul: Love Incantations of the Oklahoma Cherokee* (Dallas: Southern Methodist Univ. Press, 1982), 48–50. For further reference, see Elisabeth Tooker, ed., *Native American Spirituality of the Eastern Woodlands: Sacred Myths, Dreams, Visions, Speeches, Healing Formulas, Rituals, and Ceremonials* (New York: Paulist Press, 1979).

18. John J. McCusker and Russell R. Menard, *The Economy of British America, 1607–1789* (Chapel Hill: Univ. of North Carolina Press, 1985), 118. Cited in Gleach, *Powhatan's World,* 143.

19. Gately, *Tobacco,* 66.

20. ———., 66,

21. ———., 66.

22. "Counterblaste to Tobacco" (1604; reprint, *The Workes of King James,* 1616; reprint, London: G. Putnam and Sons, 1905).

23. *Worke for Chimney Sweepers* (1602); cited in Gately, *Tobacco,* 52.

24. Gately, *Tobacco,* 71.

25. Kansaku K., Yamaura A., Kitazawa S., "Sex Differences in Lateralization Revealed in the Posterior Language Areas," *Cereb Cortex* 9 (2000): 866–72.

26. Another famed alchemist was the physician, magician, and physicist Paracelsus

(1493–1541). His wanderings took him through Germany, France, Hungary, the Nether-lands, Denmark, Sweden, and Russia. In Russia, he was taken prisoner by the Tartars and brought before the grand Cham, at whose court he became a great favorite. Finally, he accompanied the Cham's son on an embassy from China to Constantinople, the city in which the supreme secret, the universal solvent (the alkahest), was imparted to him by an Arabian adept.

In 1526, at the age of thirty-two, he reentered Germany and, at the university he had entered as a youth, took a professorship in physics, medicine, and surgery. This was a position of considerable importance, offered to him at the insistence of Erasmus and Ecolampidus; www.alchemylab.com/paracelsus.htm (25 March 2003).

27. Thomas Hariot, *A Brief and True Report of the New Founde Lande of Virginia* (1588); (New York: Dover, 1972), 12.

28. Gately, *Tobacco,* 7.

29. That is, Attention Deficit Disorder or Attention Deficit Hyperactivity Disorder, respectively. These conditions, being incurable, are not restricted to children.

30. Gately, *Tobacco,* 36.

31. ———., 35.

CHAPTER 5. *TAPACOH* / AT THE END OF THE DAY

1. The ancient Egyptians recorded the miraculous power of Queen Isis, the Goddess, who reconstituted her murdered consort, Osiris, whose body had been torn apart and the parts scattered afar. Unable to find his penis, the queen devised the Egyptian version of a *pocohaac,* and with this wand impregnated herself. The child born of this divine act of artificial insemination was Horus, the scribe of the gods.

2. A version of Donne's "Holy Sonnets" can be found online at www.Poets-Cor-ner.org (6 January 2003); www.luminarium.org/sevenlit/donne/donnebib.htm.

3. Jonathan Edwards, sermon. Enfield, Connecticut, July 8, 1741; www.ccel.org/e/edwards/sermons.html (20 June 2003).

4. Donne, "Holy Sonnets," X.

5. Woodward, *Pocahontas,* 174 n.

6. Don Giovanni Batttista Lionello, "To the Doge and Senate, 10 February, 1617," in *Calendar of State Papers Venetian,* xiv, no. 631. Cited in Mary Sullivan, *Court Masques of James I* (New York: G. P. Putnam, 1913), 103.

7. Peter Lynch, *Learn to Earn: A Beginner's Guide to the Basics of Investing and Business* (New York: Simon and Schuster, 1995), 21–91.

8. ———., 23.

9. ———., 24.

10. While it is commonly believed that the Indian people of the East Coast didn't un-derstand the concept of property ownership, a perusal of records—such as were not de-liberately destroyed, giving rise to the common misconception—reveals a number of deeds, signed by Native parties who were relinquishing control of a given property, that testify otherwise. Betty Donahue, *American Indian Texts Embedded in Works of Canonical American Literature,* doctoral dissertation, UCLA, 1998.

11. In some versions, tobacco grows from First Woman's womb, from between her legs. However, others specify that it grows from her brain. Given that in our time brains, not wombs, are considered the site of intelligence, I have opted for the latter translation.

12. Edward D. Neill, *History of the Virginia Company of London, with Letters to and from the First Colony Never Before Printed* (Albany, NY: Joel Munsell, 1869), 84–85.

13. Samuel Purchas, *Purchas, His Pilgrimes* (n.p., n.d.), vol. 4.

14. John Chamberlain to Sir Dudley Carleton, 18 January 1616, in *State Papers Domestic James I*, xc, no. 25. Cited in Mary Sullivan, *Court Masques of James I*.

15. Gately, *Tobacco*, 43–45.

16. In 1586, a tribe of Algonquin who lived a bit south of Powhatan lands near Roanoke Island took to identifying Queen Elizabeth I as *weroanskaa*. The honor was received with delight by the queen and her counselor Sir Walter Ralegh, who, on his first expedition to North America, had brought back a captive named Manteo, along with the news of her newest honorific. Iain Gately credits Manteo with teaching Englishmen to smoke pipes.

17. Gleach, *Powhatan's World*, 31.

18. Strachey, *Historie*, 60–61.

19. In Gleach, *Powhatan's World*, 31, citing Irving A. Hallowell, "Ojibway World View and Disease," in *Contributions to Anthropology: Selected Papers of Irving A. Hollowell*, comp. Raymond D. Fogelson (Univ. of Chicago Press, 1976).

20. Marc Lescarbot, "La Conversion des Sauvages qui one est baptizes in la Novvelle France, cerre annee 1610," in *The Jesuit Relations and Allied Documents: Travels and Explorations of the Jesuit Missionaries in New France, 1610–1791*, ed. Reuben Gold Thwaites (Cleveland: Burrows Brothers Co., 1896), 1:75–77. Algonquin varies from tribe to tribe and region to region just as the Germanic languages of Europe do.

21. "John Smith's Letter to Queene Anne Regarding Pocahontas" (Caleb Johnson, Mayflower Web Pages, 1997; http://www.members.aol.com/mayflo1620/pocahontas.html; 2 January 2003).

22. ———.

23. Haile, *Jamestown Narratives*, 864.

APPENDIX 1. JOHN ROLFE'S LETTER TO MASTER THOMAS DALE

1. J. Franklin Jameson, *Narratives of Early Virginia* (New York: Charles Scribner's Sons, 1907), 237–44.

APPENDIX 3. DIEGO DE MOLINA'S LETTER TO KING PHILIP OF SPAIN, 1613

1. I.e., the caravel from which Molina had incautiously gone ashore at Point Comfort.

2. This expectation came true in the course of years. The Bahama Islands belong to England, Puerto Rico belongs to the United States—a product of the Virginia settlement—and Cuba is under American influence.

3. The present English fortress at Bermuda is considered one of the strongest in the world.

4. This is a strong confirmation by the colonists of their cruel experiences under Gates and Dale.

5. Probably the wish was father to the thought.

6. I.e., the Virginia Company.

7. Jamestown Peninsula, which is not over ten leagues up the river.

8. Francis Lembri, who was proven to be an Englishman (Irish) and hanged by Dale as a traitor, when returning to England in 1616.

9. Chesapeake Bay.

10. The widest portion of the bay is not over thirteen leagues, or forty miles.

11.   The fort at Point Comfort was called "Algernourne Fort," first established in 1608 by President Percy. The Spanish here has a play upon words that cannot be translated: "a fort [*fuerte,* strong] or rather a weak."

12.   Fort Charles on Strawberry Bank in Elizabeth City, first established in 1611.

13.   Fort Henry, on the east side of Hampton River, a musket shot to the west of Fort Charles.

14.   Jamestown.

15.   At Henrico, where Dutch Gap cuts the bend of the river. Some scattered bricks still give evidence of this early settlement.

16.   Lyon Gardiner Tyler, ed., *Narratives of Early Virginia* (New York: Charles Scribner's Sons, 1907), 218–24.

Glossary

Ahone: major deity; God; one of sacred twins, grandsons of Nikomis, Ahone created most of the earth's geophysical and climactic features in ways that would favor humans and make human life ideal. Spirit imbued with goodness, beneficence. *See* Oke.

Anishnabe, anishinabeg: Western Algonquin people; Ojibwaj, Chippewa, Blackfoot living mainly around the Great Lakes and in some areas farther west, such as Montana, and in Canada, Alberta.

Apook (apooc): tobacco plant grown in Virginia; *Nicotiana rustica.*

Apowa: Dream-Vision.

Appotomac: one of the major tribal states of the Powhatan Alliance.

Capahowasick (Capahowasc): *manitowinini* name of the peninsula where the English built James Fort/Jamestown.

Chickahominy: major tribe holding territory near James Fort; loosely allied with the Powhatan *tsenacommacah,* but often acted counter to its objectives; the tribe that captured John Smith and turned him over to Opechancanough.

Glous'gap: Ancient being, wizard, manito, or "culture" hero of northeastern Algonquin.

Granny Squannit: *see* Squannit.

Hobbomak: cave or rock formation above ground that encloses a small area; *hobbomaks* are found all over the Atlantic seaboard from Florida to Nova Scotia. Unknown origins; Native people since contact with Europeans have said they were ancient.

Huskanaw: boy's initiation ceremony; remaking ceremony; rebirth ceremony. The ceremony at which John Smith was remade or "reborn" was based on *huskanaw.*

Japazaws: lesser chief of the Patawamacks who colluded with Argall to get Pocahontas aboard Argall's ship, the aptly named *Treasurer.*

Makiawisug: plural of Makiawis; Little People; associated with Granny Squannit.

Mamanantowick: combination priest-prophet-medicine-person-shaman; great leader with these qualities.

Manito (manitou, manitoo, manido): spirit people; nonphysical or paraphysical energy; force; spirit guide. A complex term whose meaning depends on the context.

Manito aki: the world or land of the spirits; implicit order.

Manitowinini: people of the manito; Algonquin people.

Mannit: singular, explicit form of *manito;* a name for the creator manito or creation spirit being. All-That-Is; the Great Mystery.

Matachanna: woman who accompanied Pocahontas to England; wife of Uttamatamakin; said to be Pocahontas's sister or half-sister and daughter of Wahunsenacawh ("Powhatan").

Mattaponi: one of the tribes belonging to the *tsenacommacah;* Pocahontas's tribe.

Midéwewin: Great Ceremony; Great Medicine Dance, or Great Medicine Way in English; a spiritual discipline and spiritual way. In its most complete form there are seven levels of mastery; most participants complete four; the final three are taboo to all but a few.

Mohegan: a north Atlantic tribe; located in what is now called Connecticut.

Moshup: the giant in "Makiawisug, The Little People," the Mohegan story about Granny Squannit.

Nansemond: one of the tribal states belonging to the Powhatan Alliance. Smith was friends with one of its *weroances* and got permission to settle on the northern portion of their lands. He was wounded, however, and returned to England. He never returned to the tidewater area.

Nantaquod: Smith's *tsenacommacah* name.

Nikomis: Grandmother; name used to refer to Full Blooming Flower (Mature Flower) after her grandsons are born. Grand midwinter feast celebrated at Werowocomoco by the whole *tsenecommacah.*

Nonner Martha: Medicine woman of the Mohegan who was called "Granny Squannit."

Oke (Oki, Okee): one of major gods the Powhatan Alliance membership believed defended and protected them; exacting, demanding, cruel and autocratic; one of the twin grandsons of Winona, Nikomis, the Woman Who Fell from the Sky, Mature Flower. According to English sources, an effigy of this deity was carried by warriors when they entered battle, or initiated contact with strangers such as the English. John Smith wrote that Oke was an idol, made of skins, stuffed with moss, and adorned with chains and copper. An effigy of Oke presided over the ceremony where Smith was remade.

Opechancanough: Wahunsenacawh's brother; Powhatan after Wahunsenacawh retired; Paumunkey Indian.

Orapaks: town well outside James Fort range; spiritual capital moved there for a time in 1608.

Pamunkey: one of the *tsenacommacah* tribes; a river the Pamunkey people lived along. Werowocomoco was on the Pamunkey River; Wahunsenacawh's mother-tribe.

Parahunt: Nansemond tribal leader; John Smith's friend.

Patawamack: village where Pocahontas was staying when Argall took her to James Fort.

Paupauwiskey: wife of Japazaws.

Pawnee: Native American nation, linguistically Algonquin; northern Plains, Montana, North Dakota.

Pocohaac (pocohack): awl, pin, needle, bodkin; pestle; penis; a possible source of the name Pocahontas.

Ponemah: Young woman in "Makiawisug, The Little People," the Mohegan story about Granny Squannit.

Potomac: river that marks the northern boundary of the *tsenacommacah;* Washington, D.C., was built near its mouth.

Powa (pawa): a kind of energy and paranormal ability that enables one to foresee events, heal the sick, human, animal, or plant (one, two or all, depending); teleport objects; locate lost people or objects; soul-walk; shape-shift; compel others, human or otherwise; produce healthy and plentiful crops; connect with the mystery, the *manito aki,* and various manito.

Powagan: spirit guide or guardian; "personal" manito.

Powhatan: People of the Dream-Vision; often said to be the name of the falls at the northern boundary of the lands of the people of the Powhatan Alliance; the name given to Wahunsenacawh, who was Chief Dreamer of the alliance during the first years of the Virginia Company's intrusion into Powhatan Alliance homelands; large river draining into Chesapeake Bay that is now known as the James River.

Powwaw (powwow): we dream together. The modern "powwow," or gathering of tribal people for traditional dances, drumming, and associated activities, derives from the Algonquin term.

Priestess, priest: as meant in pagan religions rather than Christian ones, a practitioner of magical arts more detailed and structured than shamanic arts. A priest enacts more rigorously constructed ceremony and ritual procedures; medicine woman, medicine man.

Quahog (quohog): clam found in northern Atlantic waters; shell is white with distinct vivid violet edge on outer rim. Used to make money, jewelry, and other decorative items.

Squannit: woman-god; female deity connected to earth, seasons, climate, weather cycles, birth, death, abundance, famine, plant and animal life; leader of the Little People. *Squa* or *skaa* is the word for "woman" in most northeastern Algonquin dialects.

Sunkskaa (f.), sunk (m.): leader or chief; female and male forms of the word, respectively.

Tomokin (Tomo): *see* Uttamatamakin.

Uttamatamakin (Tomo, Tomokin): priest-envoy who accompanied Pocahontas to England; met with Ralegh, Lord Northumberland, and other wise men—astrologers and "wizards," as the English identified such men.

Virginia Company: By royal charter, the name given to all English holdings in the "New World." The land was bestowed in strips about one hundred hundred miles wide and as long as the North American continent, which they thought to be narrower than it turned out to be. What is now the United States would, by this original charter, be known as Virginia. The Virginia Company, a subsidiary of the City of London Company, was divided into the Virginia Company, Jamestown, and the Virginia Company, Plymouth, when the company was reorganized in the 1620s.

Wahunsenacawh: Powhatan and *mamanantowick,* principal chief of the Powhatan Alliance; known in English as Father of the Powhatan Alliance, and as "Great King and Impire" (emperor); Pamunkey tribal member.

Wampum: name used by Algonquins north of the *tsenacommacah* for "money" or marker of exchange value. Shell of quahog clam, highly valued though plentiful, because of its beauty and durability for jewelry and other ornamentation as well as money.

Weroance (m.), weroanskaa (f.); *also* **werowance (m.), weroansqua (f.):** political leader with a degree—greater or lesser—of shamanic and/or priestly competence; male and female forms, respectively.

Werowocomoco: major town of the Powhatan Alliance; its spiritual and government center.

Windigo: dangerous winter manito, force, being, god, or presence. It kills, devours, or drives its victims mad, turning them into thieves, rapists, murderers, cannibals, and violators of sacred ways and knowledge.

Bibliography

Adams, Richard C., et al. *Legends of the Delaware Indians and Picture Writing*. Syracuse, NY: Syracuse Univ. Press, 1997.

Astrov, Margot, ed. *American Indian Prose and Poetry*. New York: Capricorn Books, 1962.

Axtell, James. *Beyond 1492: Encounters in Colonial North America*. New York: Oxford Univ. Press, 1992.

————. *Natives and Newcomers: The Cultural Origins of North America*. New York: Oxford Univ. Press, 2001.

Ballantine, Betty, and Ian Ballantine, eds. *The Native Americans: An Illustrated History*. North Dighton, MA: World Publications Group, 2001.

Barnouw, Victor, ed. *Wisconsin Chippewa Myths and Tales and Their Relation to Chippewa Life*. Madison: Univ. of Wisconsin Press, 1986.

Barth, John. *The Sot-Weed Factor*. New York: Anchor Books, 1987.

Bierhorst, John, ed. *The Mythology of North America*. New York: William Morrow, 1985.

————. *The White Deer and Other Stories Told by the Lenape*. New York: William Morrow, 1995.

Billings, Warren M. *Jamestown and the Founding of the Nation*. Gettysburg, PA: Thomas Publications, n.d.

Brandon, William. *The Last Americans: The Indian in American Culture*. New York: McGraw-Hill, 1974.

Brinton, Daniel. *The Lenape and Their Legends, with the Complete Text and Symbols of the Walum Olum*. 1885. Reprint, Lewisburg, PA, 1999.

Brooks, Jerome E. *The Mighty Leaf: The Story of Tobacco*. London: Alvin Redman, 1953.

Burland, C. A. *The Gods of Mexico*. New York: G.P. Putnam's Sons, 1967.

Burland, C. A., and Werner Forman. *Feathered Serpent and Smoking Mirror*. New York: G. P. Putnam's Sons, 1975.

Craig, Hardin, ed. *The Complete Works of Shakespeare*. Chicago: Scott, Foresman and Company, 1961.

Culin, Stewart. *Games of the North American Indians.* 1907. Reprint, New York: Dover Books, 1975.

Curtis, Natalie, ed. *The Indians' Book: Songs and Legends of the American Indians.* New York: Dover Publications, 1950. Reprint, New York: Bonanza Books, Crown Publishers, 1987.

Derounian-Stodola, Kathryn, ed. *Women's Captivity Narratives.* New York: Penguin Books, 1998.

Donahue, Betty. *American Indian Texts Embedded in Works of Canonical American Literature.* Doctoral dissertation, UCLA, 1998.

Donnell, Susan. *Pocahontas.* New York: Berkley Books, 1991.

Egloff, Keith, and Deborah Woodward. *First People: The Early Indians of Virginia.* Charlottesville: Univ. of Virginia Press, 1992.

Erdrich, Louise. *The Antelope Wife.* New York: HarperCollins, 1998.

——. *The Beet Queen.* New York: HarperCollins, 1986.

——. *The Bingo Palace.* New York: HarperCollins, 1994.

——. *Love Medicine.* New York: HarperCollins, 1984.

——. *Tales of Burning Love.* New York: HarperCollins, 1996.

——. *Tracks.* New York: HarperCollins, 1988.

Fawcett , Melissa Jayne (Tantaquidgeon). *The Lasting of the Mohegans, Part I: The Story of the Wolf People.* Uncasville, CT: Mohegan Tribe, Little People Publications, 1995.

——. *Medicine Trail: The Life and Lessons of Gladys Tantaquidgeon.* Tucson: Univ. of Arizona Press, 2000.

Fawcett, Melissa Jayne (Tantaquidgeon), and Joseph Burchac. *Makiawisug: The Gift of the Little People.* Uncasville, CT: Mohegan Tribe, Little People Publications, 1997.

Forbes, Jack D. "Weapanakamikok Language Relationships: An Introductory Study of Mutual Intelligibility Among the Powhatan, Lenape, Natic, Nanticoke, and Otchipwe Languages." Davis, CA: Native American Studies Tecumseh Center, Univ. of California, 1972.

Gately, Iain. *Tobacco: A Cultural History of How an Exotic Plant Seduced Civilization.* New York: Grove Press, 2001.

Gill, Sam D., and Irene F. Sullivan. *Dictionary of Native American Mythology.* Santa Barbara, CA: ABC-CLIO, 1992.

Gleach, Frederic W. *Powhatan's World and Colonial Virginia: A Conflict of Cultures.* Lincoln: Univ. of Nebraska Press, 1997.

Goodman, Ronald. *Lakota Star Knowledge: Studies in Lakota Stellar Theology.* Rosebud Sioux Reservation, SD: Sinte Gleska Univ. Press, 1992.

Grant, Bruce. *Concise Encyclopedia of the American Indian.* Rev. ed. New York: Random House, 1989.

Grinnell, George Bird. *Blackfoot Lodge Tales.* Lincoln: Univ. of Nebraska Press, 1962.

Haile, Edward Wright, ed. *Jamestown Narratives: Eyewitness Accounts of the Virginia Colony, The First Decades: 1607–1617.* Champlain, VA: Round House Press, 1998.

Halifax, Joan. *Shamanic Voices: A Survey of Visionary Narratives.* New York: E. P. Dutton, 1991.

Halliwell, James Orchard, ed. *Private Diary of Dr. John Dee.* 1842. Reprint, Montana: Kessinger Publishing Company, n.d.

Henry, Jeannette, and Rupert Costo. *A Thousand Years of American Indian Storytelling.* San Francisco: Indian Historian Press, 1981.

Hirschfelder, Arlene, and Paulette Molin. *Encyclopedia of Native American Religions.* New York: Checkmark Books, 2001.

Hulme, Ivor Noël. *The Virginia Adventure, Roanoke to James Towne: An Archeological and Historical Odyssey.* Charlottesville: Univ. of Virginia Press, 1994.

Hultkrantz, Åke. *The Religions of the American Indians.* Trans. Monica Setterwall. Berkeley: Univ. of California Press, 1980.

Johnson, Steven F. *Nimuock (The People): The Algonquin People of New England.* Marlborough, MA: Bliss Publishing, 1995.

Johnston, Basil. *The Manitous: The Spiritual World of the Ojibway.* New York: HarperCollins, 1995.

———. *Ojibway Ceremonies.* Lincoln: Univ. of Nebraska Press, 1990.

Kelso, William M., et al. *Jamestown Rediscovery: The Search for the 1607 James Fort.* 7 vols. Charlottesville: Univ. of Virginia Press, 1995–2001.

Kilpatrick, Jack Frederick, and Anna Gritts Kilpatrick. *Walk in Your Soul.* Dallas: Southern Methodist Univ. Press, 1982.

Kupperman, Karen Ordahl. *Indians and English: Facing Off in North America.* Ithaca, NY: Cornell Univ. Press, 2000.

———, ed. *Captain John Smith: A Select Edition of His Early Writings.* Chapel Hill: Univ. of North Carolina Press, 1988.

Lacey, Robert. *Sir Walter Ralegh.* London: Phoenix Press, 1973.

Lame Deer, John (Fire), and Richard Erdoes. *Lame Deer: Seeker of Visions.* New York: Simon & Schuster, 1972.

Landes, Ruth. *Ojibway Woman.* 1938. Reprint, New York: Columbia Univ. Press, 1977.

Larson, Charles R. *American Indian Fiction.* Albuquerque: Univ. of New Mexico Press, 1978.

Lepore, Jill. *The Name of War: King Philip's War and the Origins of American Identity.* New York: Vintage Books, 1999.

Lewis, Janet. *The Legend: The Story of Neengay, an Ojibway War Chief's Daughter, and the Irishman John Johnston, A Libretto.* Santa Barbara, CA: John Daniel, 1987.

Lynch, Peter, and John Rothchild. *Learn to Earn.* New York: Fireside Books, 1995.

Mabie, Hamilton W. *Footprints of Four Centuries.* Philadelphia: International Publishing Company, 1894.

Mann, Barbara. *Iroquoian Women: The Gantowisas* (New York: Peter Lang, 2000).

Mavor, James W. Jr., and Byrone E. Dix. *Manitou: The Sacred Landscape of New England's Native Civilization.* Rochester, VT: Inner Traditions International, 1989.

McCary, Ben C. *Indians in Seventeenth-Century Virginia.* Jamestown Booklet no. 3. Charlottesville: Univ. of Virginia Press, 1957.

Miller, Lee. *Roanoke.* New York: Arcade Press, 2000.

Milton, Giles. *Big Chief Elizabeth.* New York: Picador Press, 2000.

Mohawk Tribe. "Creation Story: A Mohawk Account." In *Akwesasne Notes Calendar.* Mohawk via Roosevelt Town, NY: Akwesasne Notes, 1982.

Momaday, N. Scott. *The Way to Rainy Mountain.* Albuquerque: Univ. of New Mexico Press, 1969.

Monroe, Jean Guard, and Ray A. Williamson. *Native American Star Myths.* Boston: Houghton Mifflin Company, 1987.

Mooney, James. *The Swimmer Manuscript: Cherokee Sacred Formulas and Medicinal Prescriptions.* Revised and completed by Frans M. Olbrechts. Washington, D.C.: Government Printing Office, Smithsonian Institution, Bureau of Ethnology, 1932.

Mossiker, Frances. *Pocahontas: The Life and the Legend.* New York: De Capo Press, 1996.

Narby, Jeremy. *The Cosmic Serpent: DNA and the Origins of Knowledge.* New York: Jeremy Tarcher/Putnam, 1998.

Norman, Howard, ed. *The Wishing Bone Cycle: Narrative Poems from the Swampy Cree Indians.* New York: Stonehill Publishing Company, 1976.

Peat, David F. *Blackfoot Physics: A Journey into the Native American Universe.* London: Fourth Estate, 1994.

Plane, Ann Marie. *Colonial Intimacies: Indian Marriage in Early New England.* Ithaca, NY: Cornell Univ. Press, 2000.

Pringle, Heather. *In Search of Ancient North America: An Archeological Journey to Forgotten Cultures.* New York: John Wiley & Sons, 1996.

Quarles, Marguerite Stuart. *Pocahontas: Bright Stream Between Two Hills.* 1939. Reprint, Richmond, VA: Dietz Press, 1967.

Rountree, Helen C. *Pocahontas's People: The Powhatan Indians of Virginia Through Four Centuries.* Norman: Univ. of Oklahoma Press, 1990.

———. *The Powhatan Indians of Virginia: Their Traditional Culture.* Norman: Univ. of Oklahoma Press, 1989.

Schoolcraft, Henry Rowe. *Algic Researches: North American Indian Folktales and Legends.* 1839. Reprint, New York: Dover Books, 1999.

Shoemaker, Nancy, ed. *Negotiators of Change: Historical Perspectives on Native American Women.* New York: Routledge, 1995.

Stoutenburgh, John, Jr. *Dictionary of the American Indian.* New York: Random House, 1990.

Sullivan, Mary. *The Court Masques of James I.* New York: G. P. Putnam's Sons, 1913.

Tilton, Robert S. *Pocahontas: The Evolution of an American Narrative.* Cambridge: Cambridge Univ. Press, 1994.

Tooker, Elisabeth, ed. *Native North American Spirituality of the Eastern Woodlands: Sacred Myths, Dreams, Visions, Speeches, Healing Formulas, Rituals, and Ceremonials.* New York: Paulist Press, 1979.

Vogel, Virgil J. *American Indian Medicine.* Norman: Univ. of Oklahoma Press, 1990.

Wahl, Jan. *Pocahontas in London.* New York: Delacorte Press, 1967.

Walker, Deward E. Jr., and David Carasco. *Witchcraft and Sorcery of the American Native Peoples.* Moscow: Univ. of Idaho Press, 1989.

Waugaman, Sandra F., and Danielle Moretti-Langholtz. *We're Still Here: Contemporary Virginia Indians Tell Their Stories.* Richmond, VA: Palari Press, 2000.

Weatherford, Jack. *Indian Givers: How American Indians Transformed the World.* New York: Crown Publishers, 1988.

———. *Native Roots: How the Indians Enriched America.* New York: Crown Publishers, 1991.

Weinstein-Farson, Laurie. *The Wampanoag.* New York: Chelsea House, 1989.

Williams, Roger. *A Key into the Language of America; or, An help to the Language of the Natives in That Part of America, Called New-England.* 1643. Reprint, Bedford, MA: Applewood Books, 1936.

Williamson, Ray A. *Living the Sky: The Cosmos of the American Indian.* Norman: Univ. of Oklahoma Press, 1989.

Wilshire, Bruce. *The Primal Roots of American Philosophy: Pragmatism, Phenomenology, and Native American Thought.* University Park, PA: Pennsylvania State Univ. Press, 2000.

Wissler, Clark, and D. C. Duvall, eds. *Mythology of the Blackfoot Indians.* Lincoln: Univ. of Nebraska Press, 1995.

Woodward, Grace Steele. *Pocahontas*. Norman: Univ. of Oklahoma Press, 1969.

Woolley, Benjamin. *The Queen's Conjurer: The Science and Magic of Dr. John Dee, Adviser to Queen Elizabeth I*. New York: Henry Holt and Company, 2001.

Yates, Frances. *The Occult Philosophy in the Elizabethan Age*. New York: Routledge, 2001.

Zapp, Ivar, and George Erikson. *Atlantis America: Navigators of the Ancient World*. Santa Barbara, CA: Adventures Unlimited Press, 1998.

Index

Page references followed by *fig* indicate an illustration.